METHODS & PRACTICES OF PHYSICAL GEOLOGY

Fourth Edition

Edited by

R. H. Konig
D. L. Zachry, Jr.

University of Arkansas
Fayetteville, Arkansas

Pearson Custom Publishing

Burgess Publishing Division

ISBN 0-8087-5693-1

Printed in the United States of America
45 18

PEARSON CUSTOM PUBLIHING
75 Arlington Street, Suite 300, Boston, MA 02116
A Pearson Education Company

ACKNOWLEDGMENTS

This laboratory manual is written for students interested in the broad aspects of physical geology. Comments of past students, laboratory instructors, and others have been taken into consideration. The combined efforts of these individuals are greatly appreciated.

Appreciation is also extended to Janet Konig who drafted the cover and to Cindy Nottenkamper for her patience during the preparation of this laboratory manual.

CHAPTERS

APPENDIXES

CHAPTER 1

INTRODUCTION TO MINERALOGY
by
RONALD H. KONIG

Introduction

Mineralogy is a branch of geology that deals specifically with minerals. Since prehistoric times man has been drawn to minerals because minerals are literally building-blocks of such things as rocks and soils that dominate the surface layers of the Earth. Iron, used for weapons and construction; gold, silver, and platinum used for jewelry; calcium, potassium, and phosphate used as fertilizers, and other metals and nonmetals are all present in minerals that make up solid rock and soil. Despite the great value attached to minerals it is surprising that so few people have even a basic understanding of minerals and their importance to modern civilization.

Basic Properties of Minerals

Any naturally occurring inorganic, solid element or compound that has a regular internal structure and a fairly constant chemical composition is called a mineral. The internal structure and the chemical composition of minerals control many external features and are easily observed or determined in the laboratory. For example, physical properties like mineral shape, luster, specific gravity, magnetism, etc. are easily evaluated in the laboratory without the aid of expensive equipment.

Mineral Shape Minerals usually occur as irregular shaped masses of naturally occurring material in rock or soil. Occasionally a mineral will form (grow) in an environment where nothing is present to restrict its growth. Under these growth conditions the internal structure of the mineral is expressed as crystal faces, and the mineral is then referred to as a crystal. Several examples of crystals of common minerals are illustrated in figure 1-1. Crystal faces are smooth and reflect light very well. Although many crystal faces may be present on a crystal of a mineral, there are never more than two crystal faces present with the same orientation.

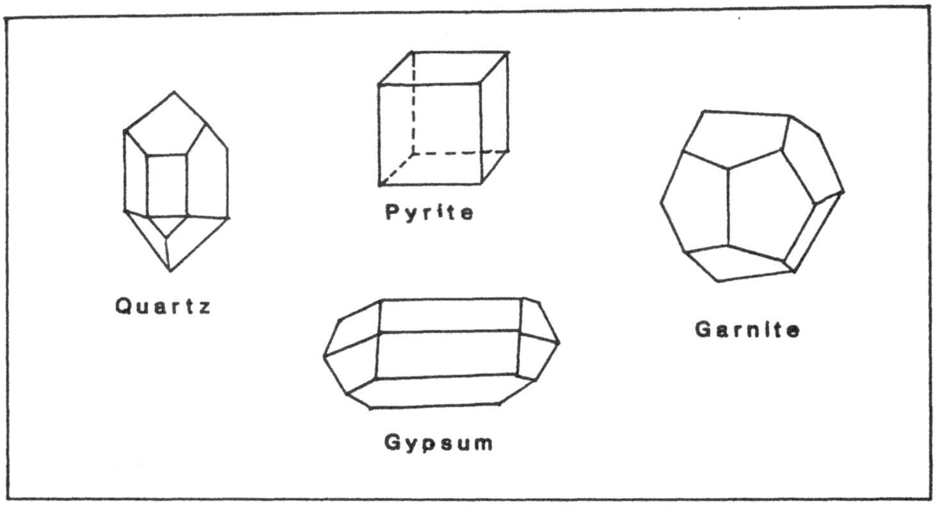

Figure 1-1. Crystal shapes of some common minerals.

Luster The appearance of a mineral in reflected light is referred to as its luster. A metallic luster is characteristic of minerals that look like metals, as for example, iron, steel, lead, aluminum, etc. In contrast, a nonmetallic luster is similar to the surface appearance of such nonmetals as glass, clay or porcelain. Metallic minerals are opaque to light, whereas nonmetallic minerals may be either transparent or translucent to light. Examples of special nonmetallic lusters include vitreous (having the luster of glass) and resinous (having the luster of amber).

Specific Gravity The weight of a mineral is easy to determine with the aid of a balance or scale. However, if we wish to determine the weight of an equal volume of one mineral and compare it to the weight of an equal volume of another mineral, a problem develops because mineral samples do not normally come in equal volumes or sizes. To overcome this volume problem, we determine the specific gravity of minerals regardless of their size and shape. The specific gravity of a mineral is a number that indicates the ratio between the weight of the mineral and the weight of an equal volume of water. If a mineral has a specific gravity of 3.0, it means that the mineral weighs 3.0 times more than an equal volume of water. The greater the specific gravity number, the heavier the mineral. Some minerals may be identified on the basis of their specific gravity because specific gravity is related to the chemical composition and the internal structure of a mineral. The specific gravity of some chemically pure minerals is given below:

Mineral Name	Specific Gravity
K-Feldspar	2.57 Light
Quartz	2.65
Garnet	4.00
Pyrrhotite	4.60
Pyrite	5.02
Magnetite	5.18
Galena	7.50
Gold	17.00 Heavy

If samples of all of the above minerals were about the same size, then lifting each of the samples would allow you to arrange them according to their specific gravity. Because minerals come in all sizes and shapes the Mineralogist measures their specific gravity with the aid of a delicate balance known as a Jolly balance and the use of a simple equation:

$$G = \frac{Wa}{Wa - Ww}$$

Where "W" equals weight, "a" equals air. "w" equals water, and "G" equals specific gravity. Thus Wa equals the weight of a mineral in air and Ww equals the weight of a mineral in water. Specific gravity and density are similar except that density requires designation of units, as for example grams per cubic centimeter.

Magnetism Minerals that are attracted to any kind of a magnet are said to be magnetic. The minerals magnetite (iron oxide) and pyrrhotite (iron sulphide) are easily identified with the aid of a simple iron magnet. However, many other minerals, especially those containing iron are attracted by powerful electromagnets. Thus some minerals may be distinguished from one another on the basis of their magnetic susceptibilities.

LABORATORY EXERCISE

1. Examine the two light colored glassy appearing samples of the mineral quartz in your tray. One sample shows crystals faces and the other does not. How many crystal faces are present on your sample? ______________________.

2. Examine the samples in your tray and decide which have a metallic luster and which have a nonmetallic luster. How many samples have a metallic luster?

   ______________________.

3. Determine which of the minerals in your tray are magnetic by using the small iron magnet provided. Do all of the magnetic minerals have the same magnetic susceptibility? ______________________.

4. Examine the samples in your tray and arrange them in order of increasing specific gravity. Have your instructor check your sample arrangement.

5. Determine the specific gravity of a metallic mineral and glassy mineral in your tray by using the balance provided. The specific gravity (G) of each sample is

   ______________________ and ______________________.

6. Why do you think your specific gravity answers do not agree exactly with those listed earlier in this laboratory exercises?______________________

   ____

   ______________________.

CHAPTER 2

IDENTIFICATION OF MINERALS
by
KENNETH F. STEELE

INTRODUCTION TO MINERAL PROPERTIES AND IDENTIFICATION

Last week we established that minerals are naturally occurring, inorganic, solid elements or compounds, that have a regular internal structure and a fairly constant chemical composition. As such they form the geologic building blocks of soil and rock in the Earth's crust. Nearly 4000 kinds of minerals are known, although only a few are volumetrically important in the composition of the earth. The most important rock forming minerals are:

1. Feldspars	5. Olivine	9. Dolomite
2. Quartz	6. Micas	10. Chlorite
3. Augite	7. Iron Oxides	11. Halite
4. Hornblende	8. Calcite	12. Clays

Other rarer minerals are important as a source of such metals as copper, iron, lead, nickel, zinc, etc. Still others have value as gemstones.

Each kind of mineral is characterized by a regular internal structure produced by an orderly arrangement of atoms and ions within a crystal network. This arrangement is constant regardless of where or how the mineral specimen was formed and is an extremely important diagnostic feature of the mineral.

Each kind of mineral is also characterized by a fairly constant chemical composition. This is determined by the elements that form the mineral. Halite (table salt) is a common mineral that is used to illustrate the importance of internal structure and chemical composition (see below).

CHEMICAL COMPOSITION

Halite is composed of the elements chlorine (Cl) and sodium (Na). Half of the ions in the crystal are chloride, and half are sodium. This ratio (50:50) is fixed within fairly definite limits for the mineral halite and constitutes its chemical definition. Other minerals may be composed of many elements in far more complex ratios, but the elements and ratios for the specific minerals are fairly constant and provide a way to chemically finger print the mineral for identification purposes .

INTERNAL STRUCTURE

The sodium and chloride ions are arranged in a particular way in the crystal structure of halite. This arrangement is illustrated in Figure 2-1. Sodium and chloride ions alternate within the crystal to produce a cubic framework. All specimens of halite are structured in this way, regardless of their place or mode of origin, and this structure is critical in identifying the mineral. Other minerals may have infinitely more complex structural arrangements of their atoms and ions, but each mineral is characterized by its own specific structural arrangement.

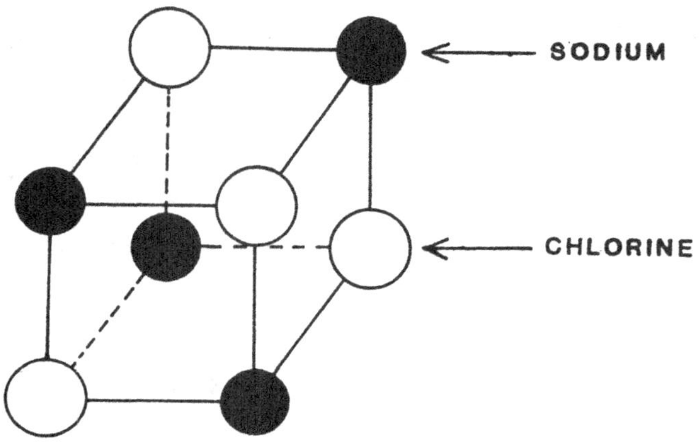

Figure 2-1. Crystal structure of halite.

PHYSICAL PROPERTIES

The internal structure and chemical composition of a mineral specimen cannot be readily determined by direct observation because atoms and ions are not visible. However, the chemical and internal structural characteristics of a mineral do determine external properties that can be observed. These properties form a basis for identifying the mineral in hand specimen without recourse to more sophisticated and costly laboratory techniques.

SPECIFIC CHEMICAL AND PHYSICAL PROPERTIES

Luster, as discussed last week, is the appearance of the surface of a mineral in reflected light. Minerals are divided into two major groups on the basis of their luster, metallic (luster similar to that of a metal) and nonmetallic.

Hardness is the resistance of a mineral to scratching. A standard scale of relative hardness was established in 1824 by Friedrich Mohs, an Austrian mineralogist, and the hardness of a mineral can be determined by comparing it to the scale. Mohs' scale of hardness is:

1. Talc (soft)	6. Potassium Feldspar
2. Gypsum	7. Quartz
3. Calcite	8. Topaz
4. Fluorite	9. Corundum
5. Apatite	10. Diamond (hard)

Talc, having a hardness of one, is very soft. Diamond, with a value of 10 is the hardest of known natural substances. On an absolute hardness scale, diamond actually is 42 times harder than talc. Unfortunately, the minerals that Mohs chose for his hardness scale are not always available to use. Because of this, common and available materials that are known to have certain values on the standard hardness scale are used. Materials generally used for the practical hardness scale in this laboratory are:

Fingernail	-	hardness 2.2 (approx.)
New Penny	-	hardness 3.2
Nail	-	hardness 5.0
Glass	-	hardness 5.5
Streak plate	-	hardness 7.0

A mineral that can be scratched with a fingernail has a hardness of less than 2.2. Specimens that can be scratched by a penny but are not scratched with a fingernail have a hardness from 2.2 to 3.2. Specimens that can be scratched with a nail, but will not scratch glass have a hardness less than 5.5, but greater than 5.0. Minerals that scratch glass, but do not scratch the streak plate, have a hardness value of 5.5 to 7.0. Finally, minerals that scratch a streak plate have a hardness greater than 7.0. A single mineral specimen can, in this way, be systematically tested to determine the range in which its hardness value falls. Precise hardness values cannot be measured using these tools, but the range of values determined provides an important clue about the mineral's identity.

Streak is the color of the ground-up or powered mineral. It is determined by scratching a specimen on a streak plate (unglazed porcelain) and observing the color of the powder. This color may be similar to that of the mineral, or it may be quite different. The color of the streak is usually more constant for a particular mineral than the color of the mineral as seen in the bulk specimen, because grain size usually affects color.

> NOTE: Because the streak plate has a hardness of about 7.0, a mineral with a hardness of 7.0 or higher will not produce a streak. Instead, the mineral will scratch the unglazed porcelain of the streak plate producing a white powder composed of porcelain.

Color is a striking characteristic of minerals and for some is diagnostic. However, several specimens of the same kind of mineral may possess different colors. Color should be considered in mineral identification, but should seldom be used as a major identifying characteristic. Color is particularly useful in identifying metallic minerals, and a few varieties of non-metallic minerals, like quartz (smokey, rose, milky, and clear quartz).

Reaction to acid is a characteristic of a few important rock forming minerals. For example, the mineral calcite ($CaCO_3$) effervesces or bubbles strongly when a drop of dilute hydrochloric acid is applied to its surface due to the production of carbon dioxide gas as follows:

$CaCO_3$	+	2HCL	——>	$CaC1_2$	+	H_20	+	CO_2
calcite	+	hydrochloric acid	——>	calcium chloride	+	water	+	carbon dioxide

Another mineral, dolomite, with a composition of CaMg $(CO_3)_2$, is distinguished from calcite by a more feeble reaction to the same acid. In order to see the effervescence associated with dolomite, it must first be powdered, and then, the acid applied directly to the powdered mineral. The reaction is now very visible. Some metallic minerals that contain sulphur will react with acid to form hydrogen sulphide gas (H_2S) that smells like rotten eggs.

Cleavage and fracture refer to the way the mineral breaks. Certain minerals, because of their specific internal structure, possess planes or surfaces of weakness along wich they tend to break. Breakage along these planes is termed cleavage and leaves a smooth surface that strongly reflects light. Minerals that break along numerous parallel planes have one direction of cleavage. Where two distinct groups of planes exist and the planes of each group are not parallel, two directions of cleavage are present. Breakage along three groups or sets of planes will produce three directions of cleavage. The intersection of cleavage planes may form a right angle (90 degrees) or may be substantially greater or less than 90 degrees. The number of directions of cleavage and the angles at which they intersect are constant for a particular kind of mineral and therefore, are important diagnostic characteristics. Cleavage characteristics of many minerals are readily evident. Recognition of cleavage in other minerals is more difficult and will require careful observation and practice. Figure 2-2 illustrates the appearance of one, two, and three directions of cleavage. Four directions and six directions of cleavage are also possible.

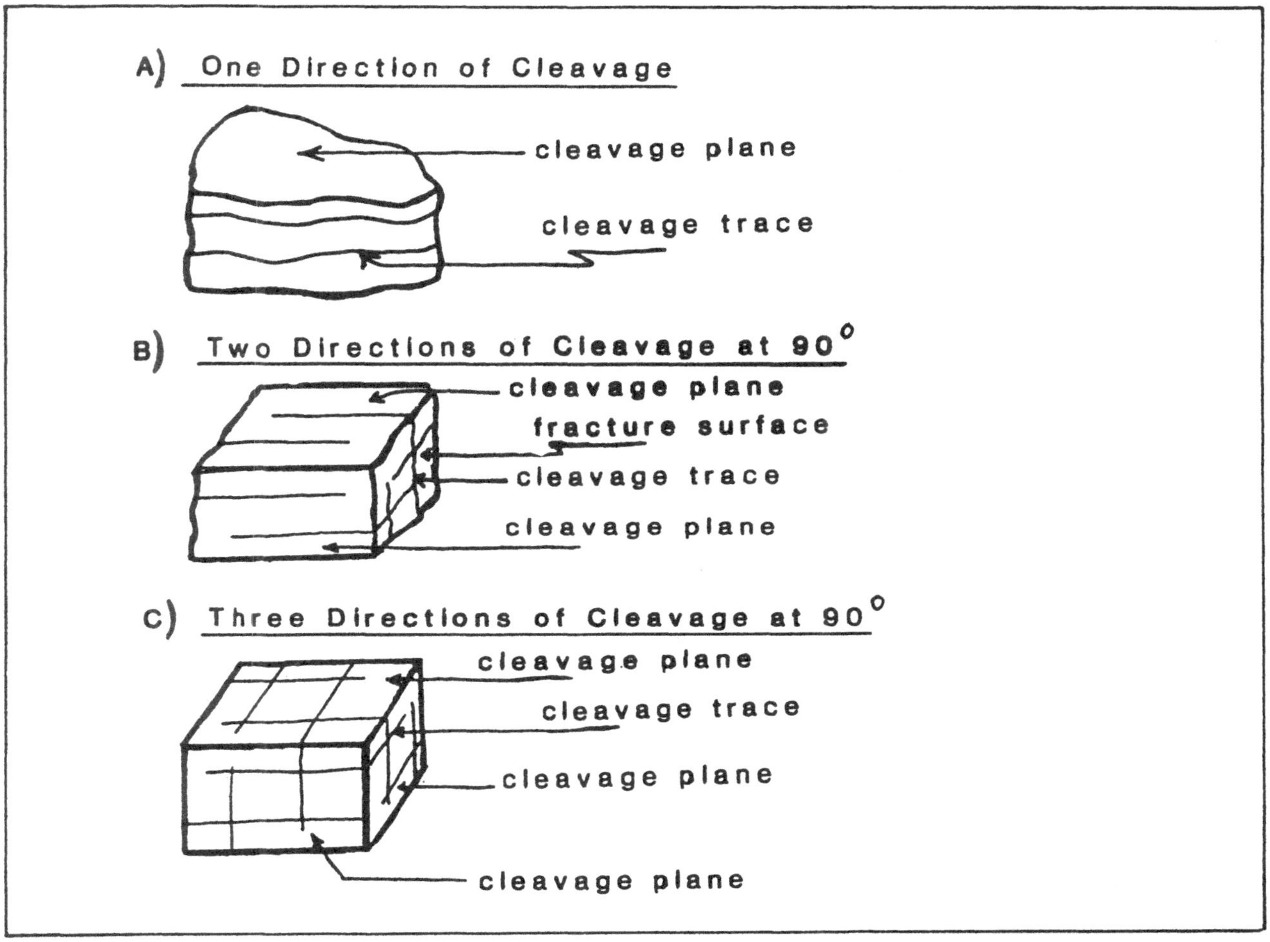

Figure 2-2. Three dimensional sketches showing cleavage planes, cleavage traces and one, two, and three directions of cleavage.

Not all minerals possess cleavage and thus break irregularly. This breakage is termed fracture. The kind of fracture displayed by some minerals is characteristic of the mineral and useful in identification. Quartz, for example, has conchoidal fracture and breaks with a pattern very similar to glass.

NOTE: Essentially all minerals are crystalline and have a well ordered internal structure. If a mineral forms without interference, such as in an open place in a rock, the mineral may develop a characteristic set of smooth crystal faces. These faces are produced by growth and not by breakage. They should not be confused with cleavage planes. Quartz crystals, may have crystals faces if the mineral formed in an open space. A quartz crystal, if broken, will simply fracture and will not break along planes parallel to the crystal faces.

Striations and thread-like markings may be seen on cleavage surfaces of some minerals. Striations are straight parallel grooves seen in reflected light on some cleavage surfaces of plagioclase feldspar. Thread-like markings are visible on some cleavage surfaces of potassium feldspar. These markings, which are due to compositional variations withing the mineral, appear as color variations just below the surface of the mineral and are neither straight nor parallel to one another and are discontinuous.

Other special properties of minerals may be important in identifying minerals. For example, thin sheets of mica are elastic, magnetite and pyrrhyotite are magnetic, certain metallic minerals (galena, magnetite, etc.) have a high specific gravity (heavy), and some evaporate minerals (halite) have a salty taste. These properties are noted in the mineral identification charts.

IDENTIFICATION OF MINERALS

Mineral identification may be accomplished by testing a specimen to determine its physical and chemical properties. All properties should be considered in that the surficial appearance of a mineral (including size and shape) may not be constant, and reliance on it alone may result in improper identification.

General Procedure

1. Determine if the luster of an unknown mineral is either metallic or nonmetallic. Minerals in the mineral identification charts are divided into these broad categories.

2. Determine the hardness range of the mineral using your fingernail, a penny, a nail, a glass plate, and a streak plate. Remember that most minerals have a rather specific hardness, but

the testing materials provide only a range of values for the mineral, not a precise value.

3. Determine the cleavage characteristics of the mineral:

   a) No cleavage
   b) One, two , three, four, or six directions
   c) Do cleavage faces intersect at right angles (90 degrees) or at larger and smaller angles?

4. Determine the color of the streak using the streak plate.

5. Test for acid reaction using dilute hydrochloric acid. If the hardness is greater than 4 on Mohs' scale, this procedure is not necessary.

6. Check for other special properties like striations, thread-like markings, specific gravity, taste, elasticity, color, and magnetism.

CHEMICAL SYMBOLS AND MINERAL COMPOSITION

The following list of elements and chemical symbols is appended for use by students who may be unfamiliar with them:

Al	Aluminum	K	Potassium
C	Carbon	Mg	Magnesium
Ca	Calcium	Na	Sodium
Cl	Chlorine	O	Oxygen
Cu	Copper	Pb	Lead
F	Fluorine	S	Sulphur
Fe	Iron	Si	Silicon
H	Hydrogen		

Chemical compounds are formed by combining one or more of the above elements to make various chemical or structural groups

Silicon and oxygen to form <u>silicates</u> (e.g. $KA1Si_3O_8$)

Carbon and oxygen to form <u>carbonates</u> (e.g. $CaCO_3$)

Sulphur and oxygen to form <u>sulphates</u> (e.g. $CaSo_4.2H_2O$)

Metals and sulphur to form <u>sulphides</u> (e.g. FeS_2)

Metals and oxygen to form <u>oxides</u> (e.g. Fe_2O_3)

Sodium and chlorine to form <u>chlorides</u> (e.g. NaCl)

Because silicon and oxygen make up about 75 percent of the Earths crust, silicate minerals (silicates) are the most abundant minerals in crustal rock. Valuable metallic materials like sphalerite (ore of zinc), galena, (ore of lead), etc. are very rare.

IMPORTANT USES OF SOME MINERALS

1. Chalcopyrite——————————Ore of copper
2. Fluorite——————————Flux in metallurgical operations
3. Galena——————————Ore of lead
4. Garnet——————————Abrasive; gemstone
5. Gypsum——————————Gypsum wallboard
6. Halite——————————Ice control agent, table salt
7. Hematite——————————Ore of iron
8. Kaolinite——————————Ceramics
9. Magnetite——————————Ore of iron
10. Plagioclase——————————Porcelain
11. Quartz——————————Glass; semi-precious stone
12. Sphalerite——————————Ore of Zinc

LABORATORY EXERCISE

The minerals listed and described in figures 2-3 and 2-4 are to be studied in the laboratory during the next two weeks. It is convenient to divide these minerals into two groups known as Mineral With Cleavage and Minerals without Cleavage. Because minerals that do not contain cleavage are easier to deal with, the next formal laboratory exercise will deal with that group.

MINERAL EXERCISE I (NO CLEAVAGE)

1. Select one of the unknown mineral samples provided and determine its luster, hardness, streak, color, etc.

2. Place the mineral sample on a blank piece of paper, and write its complete name below the sample. Paper is provided at the end of the chapter.

3. Repeat steps number 1 and 2 until all samples without cleavage have been identified.

4. Have your laboratory instructor check your answers for correctness.

5. If errors have been made, determine the physical and chemical properties again for each sample missed, and then refer to the identification tables for the correct name.

MINERAL EXERCISE II (WITH CLEAVAGE)

1. Remove all of the minerals studied last week (minerals without cleavage) from tray, place them on a blank piece of paper, and write the name of each below the sample. Have your laboratory instructor check your answers.

2. Select one of the new mineral samples in the tray and determine its luster, hardness range, streak, color, cleavage (number of directions and angle), and any other special properties.

3. Place the unknown sample on a blank piece of paper and write its complete name below the sample. Repeat this for each unknown.

4. Have your laboratory instructor check your work for correctness.

5. If time permits, exchange trays with another group in the laboratory and identify samples in that tray in case you have a quiz on minerals next week.

MINERAL IDENTIFICATION CHART I

Metallic Luster

HARDNESS	STREAK	COLOR	SPECIAL PROPERTIES	NAME
1.0	Black or Gray	Steel Gray	Soft, marks paper, greasy feel, luster may be dull, no cleavage. G=2.3	GRAPHITE C
2.5	Gray	Silver Gray	Excellent cleavage, 3 directions at 90 degrees. G=7.5	GALENA PbS
3.5 - 4.0	Greenish Black	Golden Yellow	May tarnish purple, no cleavage. G=3.5	CHALCOPYRITE $CuFeS_2$
4.0	Black	Brownish Bronze	Color and weak magnetism. G=4.6	PYRRHOTITE FeS
5.0 - 5.5	Yellow Brown	Yellow Brown Black	Structureless or radial fiberous masses, no cleavage. G=3.7	GOETHITE FeO(OH)
5.0 - 6.0	Reddish Brown	Black to Dark Brown	Granular, fiberous or micaceous, no cleavage. G=5.2	HEMATITE $Fe_3 0_3$
6.0	Black	Black	Strongly magnetic G=5.2	MAGNETITE Fe_3O_4
6.0 - 6.5	Greenish Black	Brass Yellow	Common as cubic crystals and granular aggregates, no cleavage. G=5.0	PYRITE FeS_2 (Fool's Gold)

Figure 2-3. Mineral identification chart for minerals that have a metallic luster.

MINERAL IDENTIFICATION CHART II

Nonmetallic Luster

HARDNESS	CLEAVAGE	STREAK	SPECIAL PROPERTIES	NAME
1.0	1 direction poorly developed, forms thin scales	White	White to green, soapy or greasy feel	TALC Complex Silicate
1.0 - 2.0	No cleavage visible	White	White, earthy odor, plastic when wet	KAOLINITE Complex Silicate
1.5 - 3.5	No cleavage visible	Yellow-Brown	Yellowish to dark brown, compact earthy mass	LIMONITE Hydrous Iron Oxide
2.0	No cleavage visible	Reddish Brown	Red, earthy appearance (Oolitic)	HEMATITE Fe_2O_3
2.0	No cleavage	Greenish	Green, greasy feel	SERPENTINE Complex Silicate
2.0	Perfect in 1 direction. Poor in 2 others.	White	Transparent to white, sometimes fiberous	GYPSUM $CaSO_4.H_2O$
2.0 - 2.5	Good in 1 direction	Pale Green	Green color and flaky cleavage	CHLORITE Complex Silicate
2.0 - 2.5	Perfect cleavage in 1 direction	White	Transparent in thin sheets, elastic	MUSCOVITE Complex Silicate
2.0 - 2.5	Perfect cleavage in 3 directions @ 90°	White	White to colorless, soluble in water, salty taste	HALITE NaCl
2.5 - 3.0	Perfect cleavage in 1 direction	White	Black to brown, thin elastic sheets	BIOTITE Complex Silicate
3.0 - 4.0	3 directions not @ 90°	Light Colored	Feeble acid reaction	DOLOMITE $CaMg\ (CO_3)_2$

3.0	Perfect, 3 direction not @ 90°	White	Usually white, effervesces in cold hydrochloric acid	CALCITE $CaCO_3$
3.5 - 4.0	None	Blue	Azure blue color	AZURITE Copper Carbonate
3.5 - 4.0	No cleavage visible	Green	Bright green color	MALACHITE Copper Carbonate
3.5 - 4.0	Good, directions	White	Yellowish-brown, resinous luster	SPHALERITE ZnS
4.0	Good, 4 directions	White	Colorless, yellow, Blue, green, or violet	FLUORITE CaF_2
6.0	Good, 2 direction @ nearly 90°	Pale Green	Black to dark green	AUGITE Complex Silicate
6.0	Good, 2 direction @ 60 to 120°	Pale Green	Black to dark green or brown	HORNBLENDE Complex Silicate
6.0 - 6.5	Good, 2 direction @ 90°	White	Pink, white, or green,	POTASSIUM FELDSPAR $KA1Si_3O_8$
6.0 - 6.5	Good, 2 direction @ 87 and 93°	White	White, blue gray striations on some cleavage surfaces	PLAGIOCLASE FELDSPAR $(NaCa)AlSi_3O_8$
6.5 - 7.0	No cleavage	None	Olive green, small glassy grains	OLIVINE $(FeMg)_2SiO_4$
7.0	No cleavage	None	Transparent, milky, smokey, light pink (rose) clear	QUARTZ SiO_2
7.0 - 7.5	No cleavage	None	Reddish brown, glassy grains	GARNET Complex Silicate

Figure 2-4. Mineral identification chart for minerals with a nonmetallic luster.

CHAPTER 3

IDENTIFICATION OF IGNEOUS ROCKS
by
RONALD H. KONIG

MODE OF FORMATION AND TEXTURE

A rock is a naturally formed consolidated material composed of grains of one or more minerals. Igneous, sedimentary, and metamorphic rocks form the three classes of rock that comprise the Earth's crust. Igneous rocks are those that form as magma or lava cools and solidifies. The material making up this molten rock usually comes from deep within the Earth's crust or the uppermost regions of the mantle. The term magma applies to molten rock underground. The term lava applies to molten rock at the surface. Lava may flow or erupt onto the surface through a volcano or simply flow out of cracks in the ground called fissures.

These two regions of cooling and solidification have a strong influence on the rate at which the molten rock cools and, hence, on the size of the minerals forming in the rock as the magma solidifies. Thus, lavas at the Earth's surface that comes in contact with a much cooler atmosphere or ocean water, solidify rapidly and develops small diameter minerals (fine grains). The magmas that do not rise completely to the surface, but remain fairly deep in the crust, cool more slowly due to the higher temperature of the rocks surrounding them at depth and contain larger diameter minerals (coarse grains). In this way, two different grain sizes form in the resulting igneous rock. To determine the grain size of a rock, you look at the diameter of the individual grains in the rock. The rapidly cooling lava, called extrusive because it extrudes onto the Earth's surface, forms rock that is fine-grained. Magmas that cool slowly and solidify at depth in the crust, called intrusive because they intrude into the crustal rocks, result in rock that is coarse-grained.

The grains in these igneous rocks are minerals that form as the magma cools. The size of the minerals depends largely upon how long a time they are given to form in the cooling liquid. If the lava cools quickly, the minerals have little time to grow and are usually too small to identify. Hence, the rock has an aphanitic texture (from the Greek "aphanes" meaning invisible). Intrusives cool over long periods of time, and so the minerals can grow large. The resulting minerals can be seen easily, and the resulting texture is phaneritic (from the Greek "phaneros" meaning visible). If the magma cools instantly, as when it is extruded into cold water, no minerals would have time to form and a glass would result. The texture of such a rock is called glassy. Some magmas begin to crystallize at depth and then are erupted onto the surface (or are emplaced near the surface) and hence, contain minerals characteristic of slow cooling surrounded by smaller minerals resulting from rapid cooling. The smaller minerals constitute the groundmass of the rock, and the much larger minerals are known as phenocrysts, and the resulting texture is called porphyritic. Many rocks with a two-stage cooling history are porphyritic with an aphanitic groundmass. Others may be porphyritic with a phaneritic groundmass if the minerals in the groundmass grew to a larger diameter because of a slower rate of solidification (crystallization). A porphyritic texture is particularly common in many volcanic rocks. Figure 3-1 illustrates some of the textures characteristic of igneous rocks.

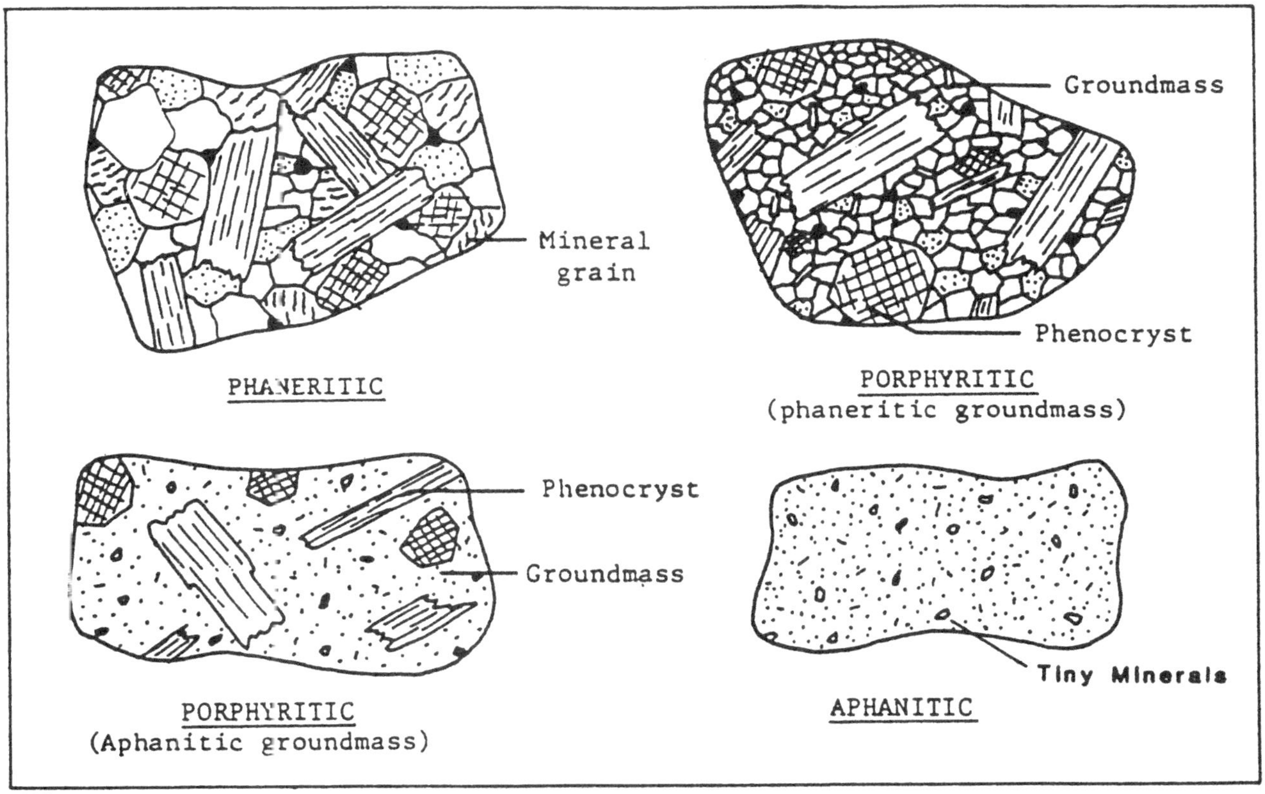

Figure 3-1. Some typical textures of igneous rocks.

Special classes of volcanic rocks may involve rapid cooling (glassy texture), trapping of gases in the molten material forming many small bubbles in the resulting rock (vesicular texture), and fracturing due to violent volcanic explosion. We refer to rock with a conspicuously glassy texture as Obsidian. In contrast, rock with many vesicles (gas cavities) and an inconspicuous glassy texture is named either Pumice, if light colored with smaller vesicles, or Scoria, if dark colored with larger vesicles. Finally, any rock type may be extensively fractured and erupted violently into the air, eventually falling back onto the ground producing layers of rock with a pyroclastic texture (from "pyro", meaning heated and "clastic", meaning fragments). This represents much of the volcanic rock in the Earth's crust. These rocks are often layered and contain angular fragments of volcanic material that become packed or welded together to form solid rocks like Volcanic Tuff or Volcanic Breccia. These rock types are difficult to recognize in hand samples.

COMPOSITION AND REACTION SERIES

Magmas and lavas are composed of various amounts of silicon, aluminum, iron, sodium, potassium, calcium, magnesium, oxygen, and minute amounts of nearly every other naturally occurring element.

Because the amount of oxygen in a magma far exceeds the amount of other elements present, the molecules that form in the liquid state of the magma are primarily oxides such as SiO_2, Al_2O_3, FeO, etc. These molecules combine to produce the minerals of which igneous rocks are composed. There are seven minerals which generally occur in igneous rocks, although all igneous rocks do not contain all of these minerals. These minerals are olivine, augite, hornblende, biotite, plagioclase feldspar, potassium feldspar and quartz.

The minerals toward the beginning of the list are deficient in silica (SiO_2), whereas the minerals at the end are silica-rich. Similarly, minerals at the beginning contain more calcium, magnesium, and iron but less sodium and potassium than those minerals at the end of the list. Igneous rocks are classified and named according to the minerals in them and the relative proportions of each mineral. Fortunately for geologists minerals crystallize in a particular order. Minerals with little silica, such as olivine and calcium-rich plagioclase feldspar, crystallize first at high temperatures. Minerals with abundant silica, such as potassium-rich feldspar and quartz, crystallize last at lower temperatures. This crystallization sequence is outlined on the next few pages:

I. Olivine is the first mineral to crystallize from a magma. This means, that with respect to the composition of the magma, higher temperatures are right for olivine to form; i.e. olivine is stable at these very high temperatures. As the magma continues to cool, much of the olivine that has formed becomes unstable and reacts with the liquid portion of the magma to form the new mineral augite. The unreacted olivine settles to form an olivine-rich magma below. In the same manner, the temperature of the magma may eventually fall below the stability range of augite, and much of it will react to form hornblende. This hornblende then reacts, as before, to form the mineral biotite mica.

These four minerals are the ones formed during the discontinuous portion of the reaction series (figures 3-2) and are commonly lumped together under the name mafics or ferromagnesium minerals, because they are all dark colored minerals rich in iron, magnesium, and/or calcium. These minerals produce dark colored igneous rocks if they are abundant in the rock.

II. In the same temperature range for which olivine is stable, a plagioclase feldspar also becomes stable. This feldspar has a very high calcium and correspondingly low sodium content. When the magma cools as with the discontinuous series, this calcium-rich plagioclase becomes unstable, but rather than reacting to form a completely new mineral, a different plagioclase feldspar is formed which has less calcium and more sodium than the first feldspar. This type or calcium loss - sodium gain reaction continues as the magma cools until the plagioclase feldspar that finally becomes stable is predominantly sodium-rich.

The calcium which was being taken from the original high temperature plagioclase has become part of the lower temperature mafic minerals, principally hornblende. Since there is no actual breaking down or change of the basic mineral type in this series, it is called the continuous reaction series.

III. After the final sodium-rich plagioclase and some biotite have formed, the remaining liquid, if there is any, crystallizes quickly and without reaction with the rest of the crystals to form potassium feldspar, muscovite, and quartz. Figure 3-2 illustrates diagrammatically the mineral relationships described above.

The reaction series just described does not account for all of the minerals in an igneous rock, but those that have been covered here usually make up the bulk of the rock. Realize that the minerals which formed last in the reaction series will do so only if there is some liquid left in the magma after biotite and sodium-rich plagioclase have formed.

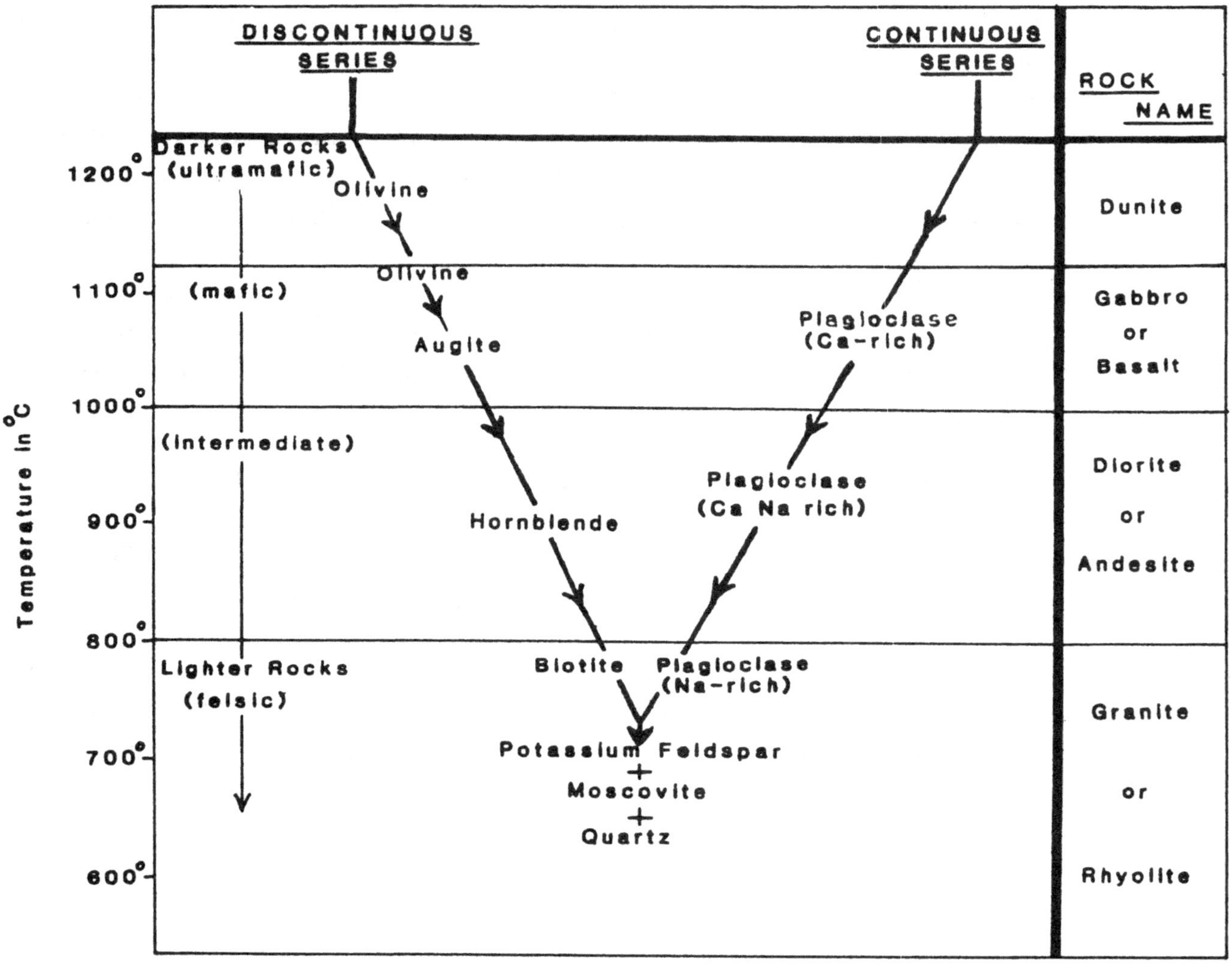

Figure 3-2. Mineral reaction series for igneous rocks. After N.L. Bowen.

Most of the last formed minerals of the reaction series are light colored and are commonly lumped together under the name felsic. These minerals are rich in silicon, sodium, and potassium and poor in iron, magnesium, and calcium. Minerals with these chemical characteristics produce light-colored igneous rocks.

If the rock we wish to study is phaneritic, and if quartz, and sodium-rich feldspar are the dominate minerals in the rock, we can say that the original magma had a felsic composition with lots of silicon, potassium, and sodium. If the rock is predominantly composed of dark minerals plus calcium-rich plagioclase, the original magma had a mafic composition with much iron, calcium, and magnesium. Rock that has a composition between that of a felsic and a mafic is classified as intermediate. Finally, a dark colored phaneritic rock composed mostly of olivine and some augite was derived from a magma of ultramafic composition that was rich in iron and magnesium.

If the rock we are studying is aphanitic, then there is no way we can actually determine what minerals are present, but a rule of thumb regarding this is; a very dark colored aphanitic rock is probably basalt, a medium-gray to brown rock is probably andesite, and a light colored to pinkish rock is probably a rhyolite. The phaneritic equivalents to the rocks named above are, respectively, gabbro, diorite, and granite.

WHERE IGNEOUS ROCKS FORM

Following the Theory of Plate Tectonics, the lithosphere has been broken into at least 13 major segments known as plates. These plates formed and may be moved by the slow rotation of convection cells operating beneath the lithosphere. Both earthquake and volcanic activity is largely confined to the contact zones separating various plates from each other.

Spreading Boundaries Igneous activity is largely confined to boundary zones between lithospheric plates. Spreading boundaries and their associated rift valleys, are flooded with mafic lava that crystallizes to produce basalts. Beneath the basaltic volcanic rock, coarser grained gabbro forms in response to a slower rate of cooling of the mafic magma. Peridotite and dunite are present at even greater depths in the asthenosphere. As these plates move laterally away from each other, more and more new basalt and gabbro, with a specific gravity of around 3.0, is formed and becomes a part of the lithosphere. Figure 3-3 shows the relationship between theoretical convection cells, oceanic lithosphere, a rift zone and the formation of basalt and gabbro in the lithosphere.

Converging Boundaries Where lithospheric boundaries are converging in response to subjacent convection cell activity a subduction zone is formed. Heavier oceanic lithospheric rock (specific gravity of 3.0) collides with lighter continental lithospheric rock (specific gravity 2.7)and the heavier basalts and gabbros of the oceanic lithosphere plunge (subduct) beneath the lighter rocks. Clearly earthquake activity would occur along this boundary zone. At greater depths, this descending basalt and gabbro will begin to melt to produce a new mafic magma. This new magma becomes less mafic and more felsic (richer in silicon, sodium and potassium) as it rises, comes in contact with, and partially melts the lighter more felsic rocks of the continental crust. In this manner both intrusive and extrusive rock material may form in association with a converging boundary. This new rock may range from gabbro to diorite to granite if intrusive, and basalt to andesite to rhyolite if extrusive. Major mountain ranges composed of composite volcanos and deformed (folded and faulted) continental rock are part of this zone of convergence. Many of these deformed rocks have undergone metamorphism due to intense pressures, elevated temperatures, and circulating fluids. New sedimentary rock may form in oceanic trenches developed along a converging plate boundary. Figure 3-3 shows the relationship between a converging boundary, the formation of a wide subduction zone (as seen when projected to the surface), and an adjacent spreading boundary.

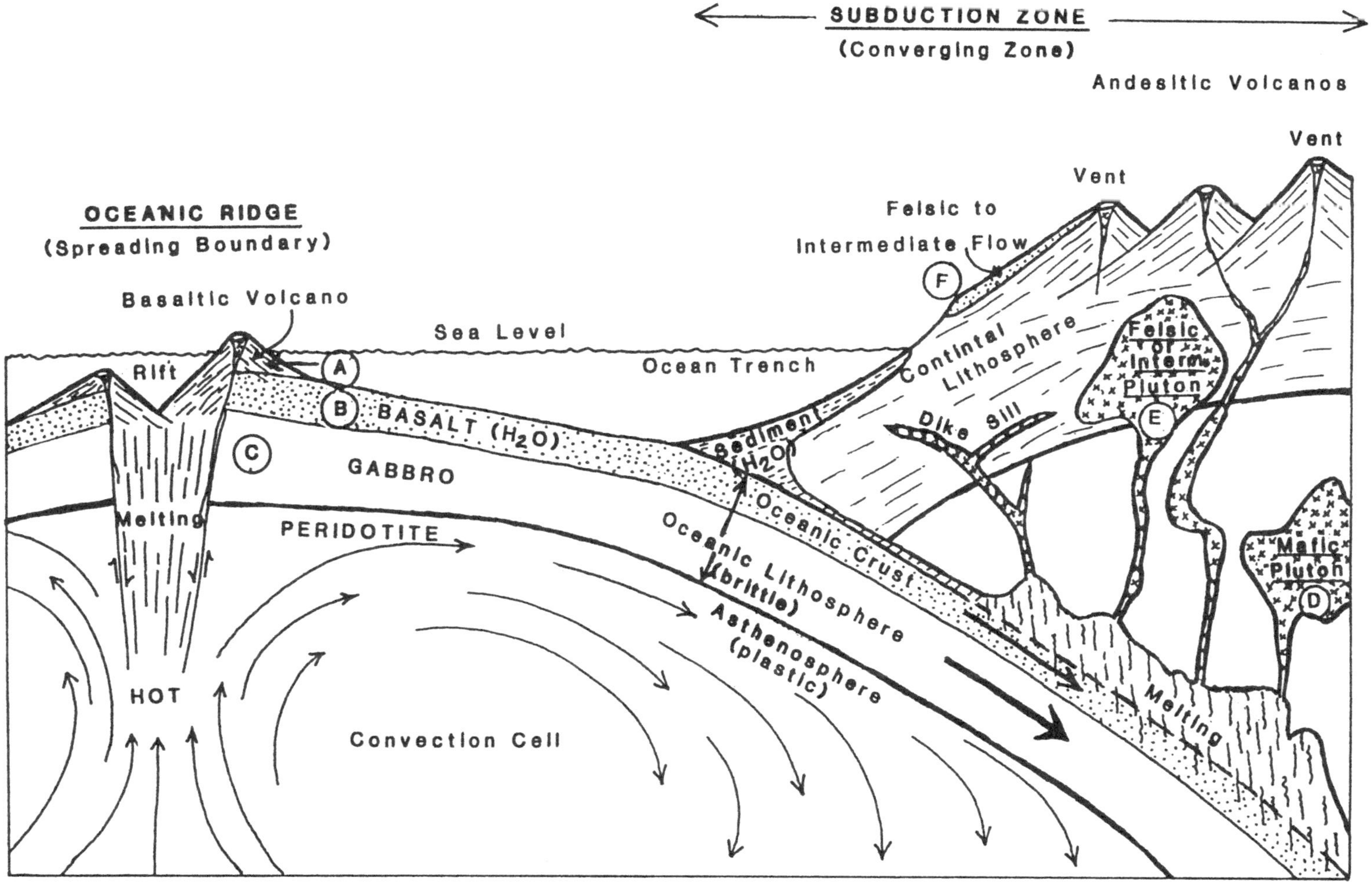

Figure 3-3. Generalized diagram showing a spreading boundary (left side) and a converging boundary (right side). Letters indicate the places of origin for various types of igneous rock.

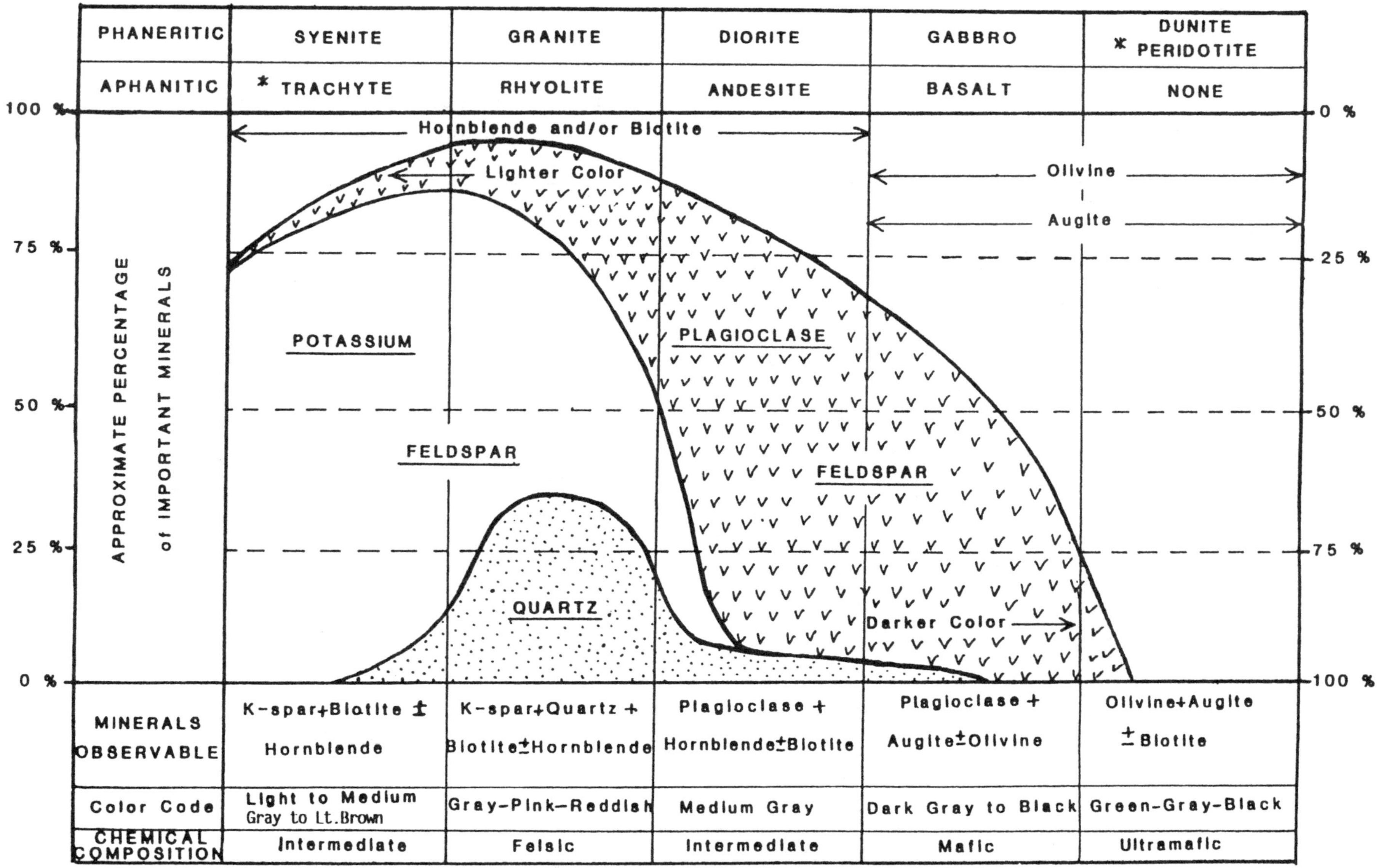

Figure 3-4. Classification chart for igneous rocks with a phaneritic or aphanitic texture. Rocks with an * are not in rock and mineral tray.

STUDY OF IGNEOUS ROCKS

As we study igneous rocks in the laboratory, three criteria will be used to name them and to draw up a hypothesis for each one about its geologic history. These criteria are: 1) texture present, 2) the relative abundance of the minerals, plagioclase feldspar, potassium feldspar, quartz, olivine, augite, and hornblende in each rock, and 3) the general color of fine grained rocks.

The study of fine-grained volcanic rock that owes its texture to rapid cooling of lava requires the use of color to determine its name. This is true because most rock names are determined by the mineral content of the rock. If rapid cooling results in a rock that is so fine grained (aphanertic texture) that minerals are not identifiable, the general color of the rock is used to determine its name. For example, a light gray to tan rock is known as a rhyolite, a medium gray to brown rock is known as an andesite, and a dark gray to black rock is known as a basalt. Even porphyritic rocks respond to this color classification if the observer ignores the color of the phenocrysts and concentrates on the color of the groundmass. Examine figures 3-1, 3-2, 3-3, 3-4, and 3-5. These figures contain all of the information necessary to discover what kind of igneous rock you may be looking at in the rock and mineral tray.

By observing the characteristics of the rocks in lab, you can construct a hypothesis as to the igneous history of each rock. For example, consider a phaneritic rock that contains much potassium feldspar and quartz and some mica and plagioclase feldspar. By consulting figure 3-4, you find that it is a granite. Knowing this, you can now relate its mineral composition to figure 3-2 and outline the region that contains the minerals you found in the rock. This gives the approximate temperature range within which the original magma cooled, and you can now construct a nearly complete history of the rock thusly:

> A magma of felsic composition was intruded into the Earth's crust at a temperature of 600 degrees C to 800 degrees C where it cooled slowly forming large diameter minerals of potassium feldspar, some sodium-rich plagioclase feldspar and quartz along with minor muscovite and biotite mica. The name of the rock is granite.

Additionally, the general place of origin of each rock type may be determined by examining figure 3-3. Figure 3-3 is a cross-sectional sketch of a portion of the earth that has been broken into two lithospheric plates following the Theory of Plate Tectonics. The letters A, B, C, D, E, and F on figure 3-3 indicate areas where specific intrusive and extrusive rocks may form. The granite intrusive described above most likely formed in the area covered by letter "E". We reach this conclusion because the rock is a Granite and has a Phaneritic (coarse grained) Texture.

<table>
<tr><td colspan="2">COMPOSITION and COLOR / TEXTURE</td><td>Felsic (light)</td><td>Intermediate</td><td>Mafic (dark)</td></tr>
<tr><td colspan="2"></td><td>K-Feldspar, Quartz, Mica</td><td>Plagioclase, Hornblende</td><td>Plagioclase, Augite, Olivine</td></tr>
<tr><td rowspan="2">PORPHYRITIC</td><td>Groundmass Phaneritic</td><td>* Porphyritic Granite</td><td>* Porphyritic Diorite</td><td>* Porphyritic Gabbro</td></tr>
<tr><td>Groundmass Aphanitic</td><td>Rhyolite Porphyry</td><td>Andesite Porphyry</td><td>Basalt Porphyry</td></tr>
<tr><td colspan="2">VESICULAR</td><td colspan="2">Pumice</td><td>Scoria</td></tr>
<tr><td colspan="2">GLASSY</td><td colspan="3">Obsidian</td></tr>
<tr><td colspan="2">PYROCLASTIC</td><td colspan="3">* Volcanic Tuff/Volcanic Breccia</td></tr>
</table>

Figure 3-5. Classification chart for igneous rocks with a porphyritic, vesicular, glassy, or pyroclastic texture. Rocks with an * are not included in your tray.

DESCRIPTIONS OF SELECTED ROCK TYPES

Syenite: A light colored, phaneritic intrusive rock composed mostly of light to medium-gray potassium feldspar and lesser amounts of black hornblende and/or biotite. Plagioclase feldspar usually is not recognizable. Quartz is not identifiable.

Granite: A light colored (occasionally reddish), phaneritic intrusive rock composed of glassy quartz, light-gray to reddish potassium feldspar, glassy plagioclase feldspar and lesser amounts of black biotite and/or hornblende.

Diorite: A medium-gray, phaneritic, intrusive rock composed of dark plagioclase feldspar, black biotite and/or hornblende and lesser mounts of gray potassium feldspar. Quartz is not identifiable.

Gabbro: A dark-gray to black, phaneritic, intrusive rock composed of dark plagioclase feldspar, dark-green olivine and black augite. Quartz is never present.

Dunite: A light to dark-green to black, phaneritic, intrusive rock composed of more than 90 percent olivine.

Periodotite: A dark-green to black, phaneritic, intrusive rock composed of a mixture of green olivine and black augite. Some biotite may be present.

Aphanitic Rocks: All aphanites are fine grained (aphanitic texture) volcanic rocks that are named with reference to their color code as given at the bottom of figure 3-4.

Porphyritic Rocks: These rocks always contain large diameter minerals (phenocrysts) surrounded by smaller diameter minerals (groundmass). Each porphyritic rock type is associated with its phaneritic equivalent by means of color. Color is determined by viewing only the groundmass. The color of the phenocrysts should be ignored.

Pumice: A light-to medium-gray volcanic rock that contains abundant vesicles (vesicular texture) that are usually smaller than the head of a pin. This rock will often float on water.

Scoria: A dark-gray to black volcanic rock that contains abundant vesicles (vesicular texture). The edges of these conspicuous vesicles are usually smooth and well rounded.

Obsidian: A glassy volcanic rock that ranges in color from pale green to reddish-brown to black. Some samples appear banded (alternating reddish-brown and black layers).

Pyroclastic Rocks: These rocks are extrusive, composed of pyroclastic material, and may be just about any color. A volcanic tuff contains rock and mineral fragments that are less than 2mm in diameter (volcanic ash) packed or welded together to form an earthy appearing, rather dense rock that may have lenses of black obsidian. A volcanic breccia is composed of rock and mineral fragments (pyroclasts) that are angular and greater than 2mm in diameter.

USES OF SOME IGNEOUS ROCK

1. Granite-----------------------Dimension stone; monuments; riprap(bank stabilization)
2. Gabbro-----------------------Dimension stone (black granite)
3. Basalt-------------------------Trap rock (bank stabilization)
4. Scoria, Pumice---------------Light weight concrete aggregate. Concrete is composed of cement, crushed stone, and water.

LABORATORY EXERCISE

1. Remove the coarse-grained (phaneritic) rocks from the tray and arrange them in a horizontal row on a piece of paper according to color (see figure 3-4).
    a) Granite: dark-gray to black minerals less than 25% and quartz is very conspicuous.
    b) Syenite: dark-gray to black minerals 25-50% and no visible quartz.
    c) Diorite: dark-gray to black minerals 50-75% and no visible quartz.
    d) Gabbro: dark-gray to black minerals 75-100%
    e) Peridotites/Dunite: dark colored and composed of either augite and/or olivine.

2. Refer to figure 3-2 and determine the temperature of formation for each of the phaneritic rocks. Write this answer next to each of the phaneritic rocks.

3. Remove all of the fine-grained (aphanitic) rocks from the tray and arrange them in a horizontal row on a piece of paper according to color (see figure 3-4).
    a) Rhyolite: white to pink to light gray
    b) Andesite: medium gray
    c) Basalt: dark gray to black

4. Remove all of the porphyritic rocks from the tray and arrange them in a horizontal row on a piece of paper according to color. Use the same color scheme as for aphanitic rocks, but do not include the color of phenocrysts in your determination. Rhyolite will always have phenocrysts of quartz regardless of its color (see figure 3-5).

5. Remove all remaining rocks from the tray that have vesicular, glassy or pyroclastic textures (see figure 3-5).
   - a) Pumice: light colored and vesicular
   - b) Scoria: dark colored and vesicular
   - c) Obsidian: red and/or black and glassy
   - d) Pyroclastic Rock: layered and with or without volcanic rock fragments. Some have lenses of obsidian.

6. Turn to figure 3-3 in lab manual. Each of your igneous rock samples may be assigned to one or more places of origin indicated by the letters A, B, C, D, E, or F on this diagram. Place the appropriate letter or letters adjacent to each rock sample so as to indicate where the sample most likely formed. Your laboratory instructor may allow this question to be completed for extra credit.

7. Examine the small slice of igneous rock on the stage of any binocular microscope in the laboratory. Note the nature of the boundary between adjacent minerals and compare them with the associated diagram.

8. Examine the thin section of igneous rock on the stage of any petrographic microscope in the laboratory. Draw a simple sketch of a small portion of your field of view in the space below and name the rock texture shown:

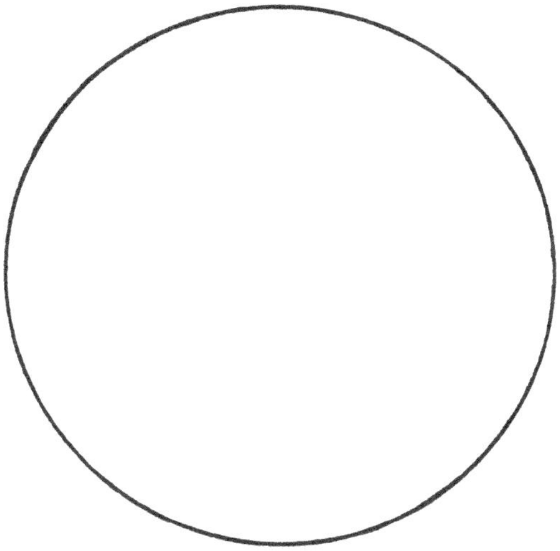

Rock Texture ___________

9. Have your lab instructor check your work for correctness and then review the minerals studied in previous labs to prepare for a possible quiz next week.

CHAPTER 4

IDENTIFICATION OF SEDIMENTARY ROCKS
by
DOY L. ZACHRY

INTRODUCTION

Sedimentary rocks are composed of materials derived by weathering of existing rocks. These materials are transported (eroded) and eventually deposited at various places on the Earth's surface to form accumulations of sediment. The accumulations are buried and eventually become consolidated (lithified) to produce sedimentary rocks.

Sedimentary materials may be derived as solid particles from a previously existing source rock and transported as solids to a place of accumulation. Particles of sand, silt, and clay carried by river currents to the sea for deposition are examples. Sedimentary rocks composed of these kinds of materials are broadly termed clastic rocks.

Weathering processes acting on source rocks may also dissolve materials and produce solutions that contain calcium, sodium, magnesium, and other soluble elements. The materials are transported in solution to places where either chemical or biologic conditions are proper for their precipitation and accumulation. Sedimentary rocks composed of these materials are broadly termed either organic rocks or chemical rocks. Today sediments are accumulating on the Great Bahama Bank across the Florida Straits from Florida. The bank is isolated from a source of clastic sediment and is covered with warm shallow water. Marine organisms flourish in the shallow water and extract calcium carbonate from solution to form calcite and aragonite shells. After the organisms die, the shells accumulate on the sea floor to form a sedimentary deposit that in time will become an organic sedimentary rock.

Water on the Great Bahama Bank is quite warm and extensive evaporation occurs. This produces high salinities and causes calcium carbonate and other compounds to precipitate as tiny crystals. These crystals accumulate on the sea floor to form sediment without the intervention of organic activity and produce chemical sedimentary rocks. Substantial volumes of sedimentary rock have been formed in the geologic past by both organic and chemical processes.

MINERAL COMPOSITION

The minerals that compose most sedimentary rocks are derived from previously existing rocks at the Earth's surface. Theoretically, any previously existing mineral could become a component of a sedimentary rock. In actual occurrences, however, a few fairly common minerals dominate the composition of most sedimentary rocks. This is because chemical weathering at the earth's surface destroys many of the unstable mineral components and leaves the stable ones to be eroded, and deposited as sediment. As a result, the mineral composition of sedimentary rocks is usually less complex than that of either igneous or metamorphic rocks. The dominant components of sedimentary rocks are:

Quartz	as sand or silt grains derived from older igneous, sedimentary, and metamorphic rocks by mechanical weathering.
Clay minerals	derived from unstable silicate minerals by chemical weathering processes. Kaolinite is an example.
Calcite	precipitated from solution by chemical and/or biological processes.
Rock fragments	particles composed of several mineral grains or crystals that were not desegregated during weathering and erosional processes.
Feldspar	derived from older igneous, metamorphic and, very rarely, sedimentary rocks. Feldspar minerals are not chemically stable and require either rapid erosion or erosion in a cold or dry climate for preservation. They are destroyed if exposed to chemical weathering processes for extended lengths of time.

Quartz, clay, feldspar, and rock fragments are important components of clastic sedimentary rocks. Calcite is precipitated from solution and is an important constituent of chemical and organic rocks. Occasionally clastic and nonclastic components may occur together in a rock. Quartz grains in sandstone may be attached to each other by calcite cement. Usually, however, either clastic or nonclastic components predominate.

CLASTIC SEDIMENTARY ROCKS

The texture of a sedimentary rock refers to the size, shape, and arrangements of component grains. Textural characteristics are very important to the understanding of the origin of sedimentary rock. They provide clues that allow deductions about the history of a particle during transportation, and about the physical conditions that existed as the particle was deposited. Texture is also an important property used in the classification of clastic rocks.

CLASTIC TEXTURE

Grain size A basic textural feature used in sedimentary rock classification is particle or grain size. A standard grain size scale is used to make this judgement (Figure 4-1).

The average grain size of a clastic rock may be generally correlated with the distance of transport of particles important in its composition. Coarse particles accumulate near the source area whereas finer particles may be transported over great distances before accumulation. Exceptions exist to this generalization, and other data is necessary for confirmation.

Grain Size Term	Definition
Gravel or Rubble (Coarse Grained)	Average grain diameter greater than 2.0mm (size of BB shot or larger). The grains can actually be measured using a metric scale or by comparison with a grain size chart.
Sand (Medium Grained)	Grain size ranges between 0.062mm and 2.0mm. In practical terms these grains range down from Pea-size to the size of fine sugar or salt. They are all visible to the naked eye.
Silt (Fine Grained)	Grain size between 0.004mm and 0.062 mm. Silt has a granular or grainy feel to the touch, but individual grains are not visible to the naked eye. These grains are all smaller than fine sugar or salt.
Clay (Very Fine Grained)	Grain size is less than 0.004 mm. The rock surface feels smooth to the touch. Grains are not visible.

Figure 4-1. Grain size scale used in classifying clastic sedimentary rocks.

Grain angularity Most grains of clastic rocks are angular when they are produced in a source area. During transport the margins of the grain are abraded and rounded. The degree of angularity a grain possesses may in many instances reflect the distance of transport the grain has been subjected to before deposition. Grain angularity also serves as a basis for dividing coarse-grained rocks (average grain size greater than 2 mm) into the two groups as follows:

1) Conglomerate——————rounded grains
2) Breccia————————angular grains

Grain sorting A measure of how equal in size the particles or grains that compose a rock are is called sorting. Rocks that contain particles of varied sizes are said to be poorly sorted. In contrast rocks composed of grains of essentially the same size are termed well sorted. The physical conditions of wave or current turbulence are capable of separating fine and coarse grains and depositing them in different places producing well sorted sediment in each place. Modern sand beaches are places of sediment accumulation. The sediment is subjected to intense wave turbulence that removes fine particles and leaves only coarse particles behind. This produces well sorted sediment that in time become well sorted rock.

The physical conditions that exist at other places of sediment accumulation that are not capable of achieving this separation produce poorly sorted sediment. Glaciers transport all sizes of sediment with equal ease. Where the glacial ice melts, sedimentary particles of all sizes accumulate together and the deposit is poorly sorted. In time poorly sorted rock is formed.

Grain sorting is not important in the classification of sedimentary rocks. However, it is important in that the sorting of a rock provides clues to the physical conditions that existed when the rock was deposited as a sediment.

Kinds of Clastic Sedimentary Rocks

Clastic sedimentary rocks are classified on the basis of mineral composition and textural properties. The clastic rock identification chart (Figure 4-2) summarizes the major kinds of sedimentary rocks along with their mineralogical and textural properties. To use the chart, one first establishes the average grain size of the rock. Use Figure 4-1 as a guide to determine if the grain size is very fine, fine, medium or coarse-grained. Next the rock is examined to determine its mineral composition.

Coarse-grained rocks with rounded particles are called conglomerates. If the particles are composed of the mineral quartz, the name quartz conglomerate is applied. Arkosic conglomerate composed of feldspar grains, and chert conglomerate composed of chert grains are other possibilities. Coarse-grained rocks with angular grains are termed breccia. Quartz, arkosic, and chert breccias occur as sedimentary rocks.

GRAIN SIZE	COMPOSITION AND SHAPE	ROCK NAME
COARSE-GRAINED (greater than 2.0 mm)	Rounded grains of quartz, feldspar or chert	Quartz Conglomerate Arkosic Conglomerate Chert Conglomerate
	Angular grains of quartz, chert or feldspar	Quartz Breccia Chert Breccia Arkosic Breccia
MEDIUM-GRAINED (between 0.062 and 2.0 mm)	Quartz Grains	Quartz Sandstone
	Feldspar Grains	Arkosic Sandstone
FINE-GRAINED (Between 0.004 and 0.062 mm)	Too fine for mineral identification. Probably quartz.	Siltstone
VERY FINE-GRAINED (less than 0.004 mm)	Too fine for mineral identification. Probably clay minerals.	Shale

Figure 4-2. Identification chart for clastic sedimentary rocks.

Medium-grained rocks are called sandstone (Figure 4-2). Because the grains are small, the angularity or roundness of the grains is not considered. The mineral composition is, however, important. Thus the name quartz sandstone is applied to rocks composed of quartz grains and arkosic to those composed of feldspar.

Fine to very fine-grained rocks are composed of particles too small to correctly determine their mineral composition. The name siltstone is applied to fine-grained rocks, and shale to very fine-grained ones. Siltstone feels granular or grainy to the touch. The grains are so small that the rock feels smooth to the touch. The grinding of a small piece of shale between your teeth does not produce the gritty sensation typical of siltstone.

ORGANIC AND CHEMICAL ROCKS

Sedimentary rocks formed through biologic activity or by direct precipitation from sea water are more difficult to categorize than clastic rocks because of their more diverse origins. Major kinds of organic and chemical rocks include limestone, dolostone, chert, coal and bauxite. Limestone for example is composed almost entirely of the mineral calcite, but the calcite may occur as fossil fragments (shells and skeletons of marine organism), as tiny spherical grains called ooliths, or a small irregular crystals of calcite. Recognition of these different grain types is important in efforts to determine the origin of the rock. A classification based on the composition of the rock, and on the constituents that compose it, is an effective way to deal with the more common organic and chemical rocks. Some of the more important constituents of organic-chemical rocks and the specific kinds of rock in which they occur include:

Calcite Grains Sedimentary rocks composed of calcite are called limestone. Many limestones contain recognizable fragments of skeletal organisms or fossils and are called fossiliferous limestone. In some occurrences the fossils may be so complete that the individual organism may be identified. In other instances the fossils are badly fragmented. The fossils may be seen but not identified. Some limestones are composed of the microscopic skeletons of tiny organism and form a rock called chalk. Most chalks have white, smooth surfaces because the skeletons are so small.

Ooliths are smooth, spherical grains of calcite similar to sand in size. Limestones in which they are the dominate constituent are called oolitic limestones.

Micrite is a limestone composed of microscopic crystals of calcite. Because the crystals are so fine, the rock has a smooth surface and breaks with a conchoidal fracture. Oolitic limestone and micrite are chemical rocks, whereas chalk and fossiliferous limestone are organic rocks. Because they are all composed of calcite, limestones effervesce with dilute hydrochloric acid. They are also easily scratched with a nail or glass plate because calcite has a hardness of 3.

Dolomite Grains Dolostone is a finely crystalline sedimentary rock composed of the mineral dolomite. The rock, of chemical origin, is similar to limestone, but does not effervesce with dilute hydrochloric acid unless powered. Powdering increases the surface area of the dolomite and changes its feeble acid reaction into a strong reaction that is easily observed.

Quartz Grains Quartz is an important mineral in clastic sedimentary rocks as visible sand grains. Microscopic crystals of quartz form chert, a rock formed by both chemical and organic processes. Most chert appears to contain the remains of microscopic organisms that used silica to construct their skeleton, just as chalk is formed from the skeletons of organisms that used calcium carbonate. Chert samples may be white, red, green or black, depending on the kinds of impurities they contain. The rocks have smooth surfaces and break with conchoidal fracture. Because chert is composed of quartz, it is very hard and will scratch a glass plate. It will not effervesce with acid like specimens of similar appearing micrite.

Carbon Coal is an organic sedimentary rock composed of various compounds of the element carbon. It is formed from the remains of plants that are buried, compressed and subjected to elevated temperatures causing them to lose most of their volatile constituents. Because the compression is severe, individual plants have lost their form and are not recognizable. The resulting material is brown or black. With less severe burial and compression a low rank, brownish coal called lignite is formed. More intense compression and slightly elevated temperatures forms higher rank, black bituminous coal. Bituminous coal has a dull luster and is soft. With very high pressures and much higher temperatures, bituminous coal changes into anthracite, a much harder metamorphic rock with a shiny luster.

Clay Minerals Clay minerals are important components of shale. These minerals may also form an important chemical sedimentary rock under certain conditions. Clay minerals contain large amounts of aluminum and silica. Under humid, tropical conditions, clay-rich rocks are intensely affected by chemical weathering processes. Silica is selectively removed forming an aluminum-rich soil called an oxisol or laterite. Laterites may in time harden through loss of water and become the chemical rock bauxite. Because this rock contains little silica it may be refined to produce the metal aluminum.

Weathering of sufficient intensity to produce bauxite from syenite occurred 35 to 50 million of years ago in Central Arkansas. The result has been the formation of economically valuable bauxite deposits from which aluminum is obtained. Thus Arkansas bauxite is an ore of aluminum.

Bauxite is a moderately soft, white, red to brown rock. The Arkansas deposits are characterized by pea-sized, spherical particles known as pisoliths that developed during the intense weathering process.

CLASSIFICATION AND IDENTIFICATION CHART

Figure 4-3 summarizes the organic and chemical rocks according to their composition and other properties. Several aspects that may make their recognition easier are itemized below:

1. Fossiliferous limestone and oolitic limestone are identified by visually recognizing the fossil fragments and ooliths. Because both are composed of calcite they effervesce with dilute hydrochloric acid.

2. Dolostone is similar in appearance to limestone, but does not effervesce in acid unless powered.

3. Shale, a clastic rock composed of clay minerals; micrite, a chemical rock composed of calcite; chalk, an organic rock also composed of quartz, are all very fine-grained and individual mineral components cannot be visually recognized. Chalk, a white rock, and micrite a gray to brown to black rock, effervesce with dilute hydrochloric acid. Shale, a platy, usually dark gray to black rock, and chert do not effervesce. Shale is soft and easily scratched with a nail, whereas chert is very hard and scratches glass.

COMPOSITION	CHARACTERISTICS	ROCK NAME
Calcite	Visible fossil fragments	FOSSILIFEROUS LIMESTONE
	Visible ooliths	OOLITIC LIMESTONE
	Microscopic fossil fragments; white In color, chalky surface	CHALK
	Microcrystalline; gray brown to Black in color	MICRITE
Dolomite	Microcrystalline; feeble acid reaction	DOLOSTONE
Quartz	Microcrystalline; very hard	CHERT
Carbon Compounds	Black, blocky; will mark paper	BITUMINOUS COAL
Clay Minerals	Red; moderately soft with pisoliths	BAUXITE
Halite	Light to medium gray; salty taste	ROCK SALT

Figure 4-3. Identification chart for rocks that originate from chemical organic and activity.

DEPOSITIONAL ENVIRONMENTS

A sedimentary rock sample can be characterized in terms of its mineral composition and texture. However, the sample is more than just an object that possesses these features. Before becoming a rock, it was composed of sediment that accumulated at a specific place and under specific conditions in the geologic past. Its characteristics, considered with our knowledge of how modern sediments accumulate, provide clues about conditions that existed in the geologic past where the sediment was deposited.

Sediment is accumulating today in a variety of places. Beaches, eolian dunes and organic reefs are examples. Each of these places is characterized by specific physical, chemical, and biological conditions, and each is termed a depositional environment. The physical, chemical and biological conditions that characterize an environment may impart to sediment specific characteristics that may be recognized in the rock long after the environment has disappeared.

A major goal of geologists is it identify and reconstruct ancient depositional environments by examining the composition, texture and sedimentary structures of rocks that were deposited as sediments in those environments. This reconstruction is very important from a historical stand point in that geologists are interested in the kinds of environment that have existed at particular times in the past. It is also important from a practical, economic viewpoint. Sediment deposited in ancient river channels is now preserved as sandstone, often buried by sedimentary rock that formed in other environments. Some of these ancient channels contain permeable sandstone that served as conduits through which uranium-bearing fluids moved and ultimately deposited uranium ore. Ancient, now-buried organic reefs often serve as reservoirs for petroleum, and their recognition is an important exploration tool in the search for this resource.

Clastic sedimentary rocks are basically affected by the physical aspects of the environment. The turbulence of the waves and the velocity of the currents in an environment determines the size of the sedimentary particles that can accumulate. In environments with swift currents or where waves are turbulent only coarse-grained sediment may be deposited. Finer grained sediment remains in suspension and is carried to quieter environments for deposition.

Chemical and organic rocks are dependent on all aspects of the environment. The chemical or biological characteristics of the environment determines the size and kinds of particles that are precipitated chemically or secreted biologically. The physical conditions of waves and current activity determines how these particles are distributed.

It is important to understand that each kind of sedimentary rock may be deposited in several different environments. Clay, which ultimately forms shale, may accumulate in any environment where currents are slow and waves are nonexistent. The deposition of gravel to ultimately form conglomerate requires an environment with strong currents. A number of environments may satisfy these requirements.

Several important depositional environments and the conditions that characterize them are described below. Their relationships to each other are illustrated in figures 4-4 and 4-5.

Clastic Depositional Environments

Alluvial Fans: Alluvial fans are aprons of sediment deposited at the foot of mountain ranges in arid climates. Streams bring the sediment from the mountains, but in the absence of rain in surrounding desert areas are unable to maintain sufficient flow to carry sediment away from the mountain front.

Alluvial fans are composed of coarse boulders and cobbles near the mountains and of finer sand and silt away from the mountain source. The sediment is poorly sorted and may contain unstable minerals like feldspar.

Eolian dunes: Eolian dunes are composed of wind-blown sand. The sand is extremely well sorted and the grains are usually rounded. Individual grains of sand may have frosted surfaces.

Alluvial sediments: Alluvial sediments are deposited by streams either directly in the channel or, during floods, on the broad flood plains that are present on either side of the channel. Channel sediment is moderately well sorted and is composed of either sand or gravel. Flood plain sediment is composed of clay and silt and is usually poorly sorted.

Marshes: Marshes are low, swampy areas located on coastal plains near the sea. They also occur on alluvial flood plains and on the emergent parts of deltas. Sediment accumulates very slowly in marshes and vegetation grows luxuriantly. These factors provide an ideal setting for the accumulation of plant remains, uncontaminated by silt or clay, and for the eventual formation of coal from this organic material known as peat.

Delta: Deltas are masses of sediment deposited by rivers at their mouths. They form seaward-facing bulges of sediment and are characterized by several depositional conditions. Sand accumulates at the seaward edge of deltas and in distributary channels. Clay, silt and sand are deposited on top of the sand as the delta builds seaward. Marshes on the emergent parts of deltas are important environments for coal formation.

Lagoons: Lagoons are bodies of water between barrier islands and the mainland. They are protected by the islands from the waves and turbulence of the open sea. Clay and occasionally silt are deposited in this quite, protected setting. The absence of plant remains distinguishes this environment from the marsh environment.

Barrier islands: Barrier islands are narrow, very elongate islands separated from the mainland by a lagoon. These islands are composed of sand deposited under the influence of wave activity. Because of this influence the sand is normally well sorted.

Shelf: The shelf is the moderately shallow sea floor that extends seaward from a barrier island or delta to the edge of the continental slope. Most of the shelf is below wave base and is not severely effected by current activity. Fine-grained sediments like silt and clay are transported here from the mainland and accumulate in the quite water. Coarser sediment like sand and gravel remain adjacent to the shoreline in more turbulent water and are not characteristic shelf deposits.

Nonclastic Depositional Environments

Oolith shoals: For their formation, ooliths require shallow, turbulent water where calcium carbonate precipitation can take place. They frequently occur in narrow shoals around the margin of shallow shelves that face the open sea. These margins are subjected to intense wave and current activity.

Skeletal sand shoals: Shells and skeletal fragments of organisms accumulate on shallow expanses of marine shelves. The water here is slightly deeper and less turbulent than the water that characterizes oolith shoals. Organic activity is necessary to produce the sedimentary particles.

Organic reefs: Organic reefs are produced by the intense growth of organism, such as corals and calcareous algae, at the margin of shallow shelves where nutrients are provided in abundance from the open sea. The reef building organisms must have strong calcareous skeletons in order to withstand the waves that attack the shelf margins.

Lagoons: Lagoons may develop between barrier reefs and the shallower parts of the shelf. They are also features of the shelf interior where slightly deeper water prevails. Lagoons are protected from wave and current turbulence by reefs or oolith shoals at the shelf margin. Very fine lime mud accumulates under these quite conditions to eventually form the rock known as micrite.

Open sea: The deep oceanic basins are not favorable sites for the production of carbonate sediment. Ooliths and most organisms require warm, shallow water. Most nonclastic sediment of the deep sea floor is composed of the shells of floating microorganism that sink to the bottom after death. The shells are very tiny and produce very fine grained sediment. Chalk composed of calcareous shells, and chert composed of siliceous shells may form in this way.

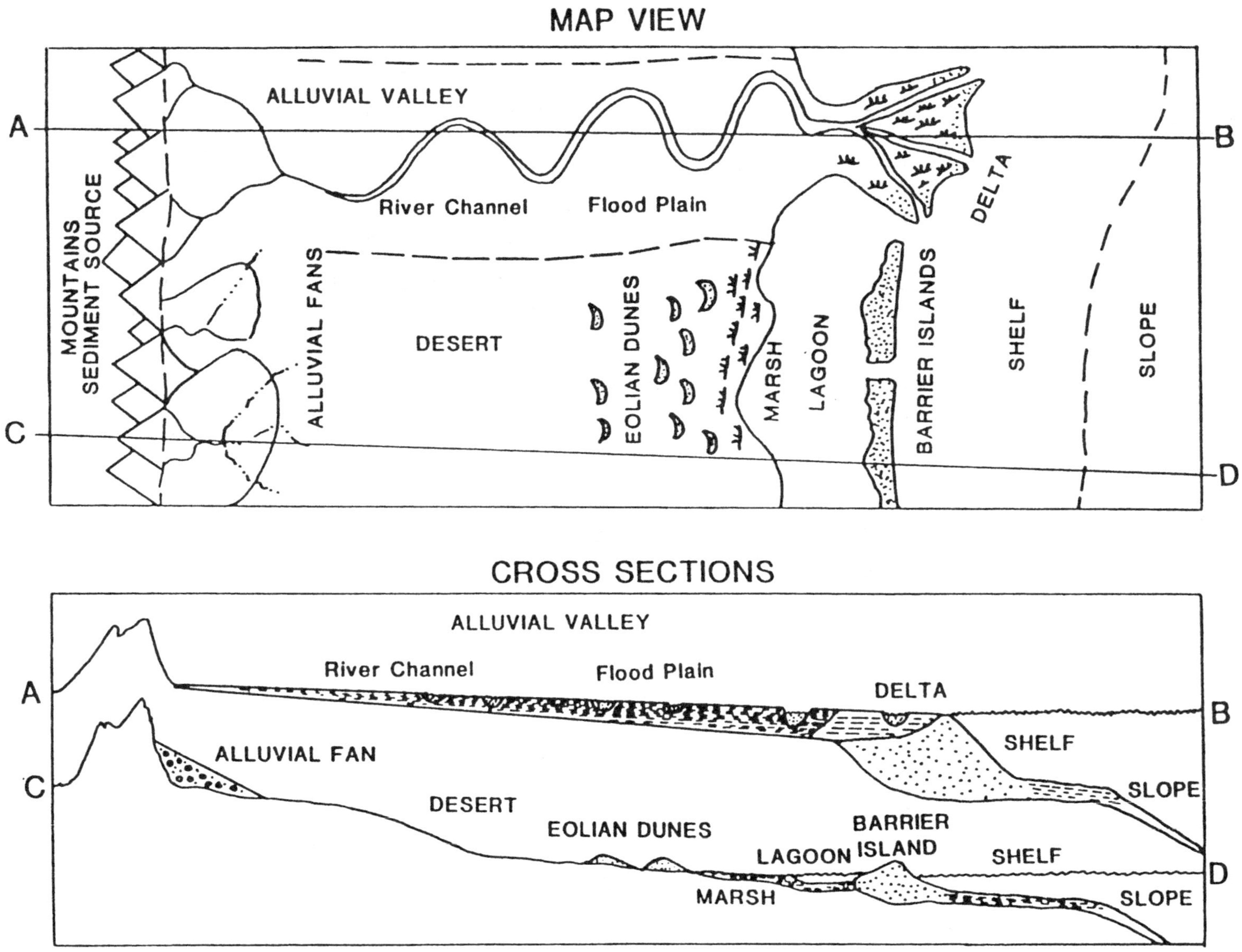

Figure 4-4. Clastic depositional environments shown in map view (above) and in cross sections (below).

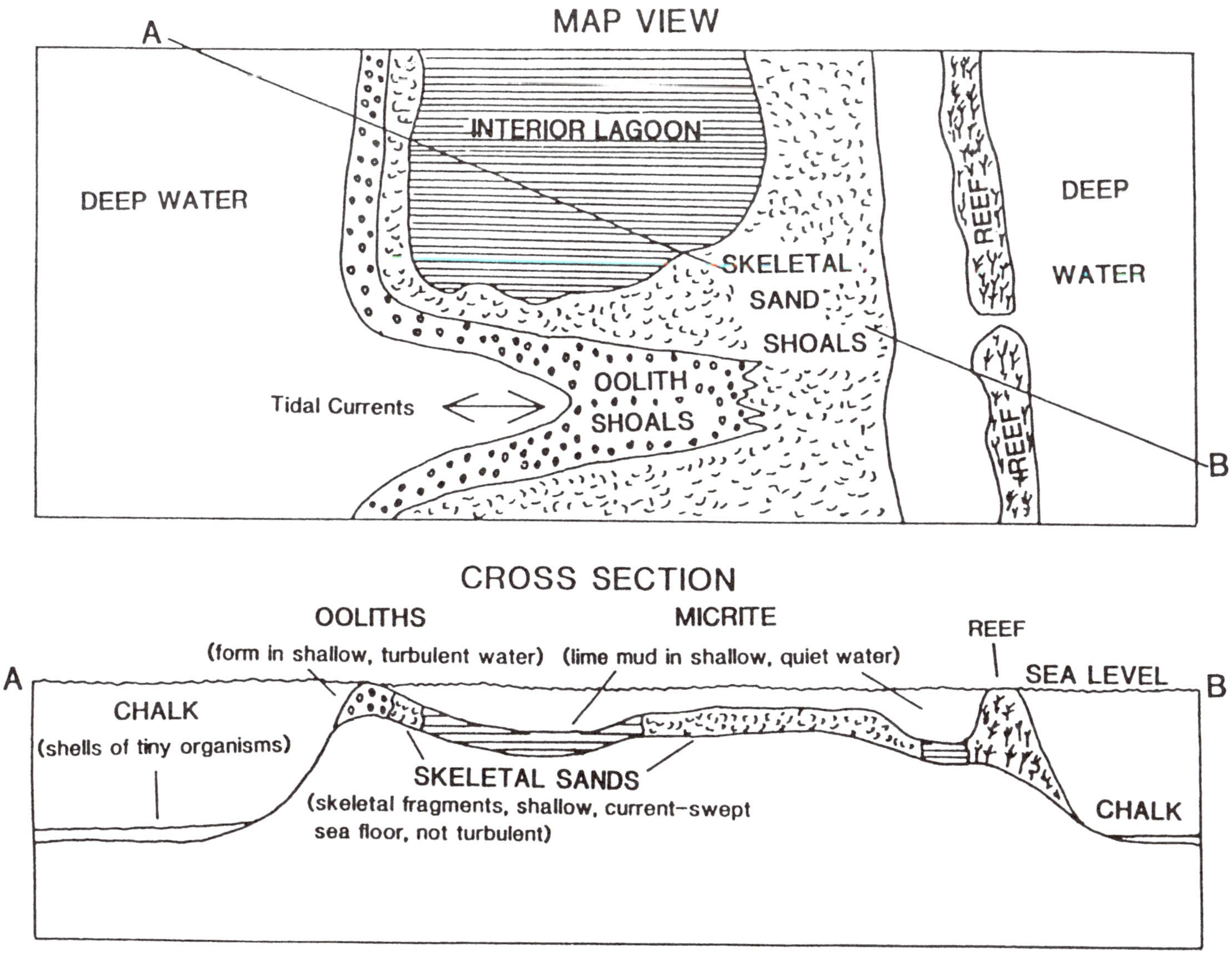

Figure 4-5. Map view (above) and cross section (below) showing various depositional environments and associated nonclastic sediment.

IMPORTANT USES OF SOME SEDIMENTARY ROCKS

1. Limestone ———————————Crushed stone for construction and concrete aggregate. Building stone. Lime manufacture. Limestone is sold as marble by industry.

2. Sandstone ———————————Building stone.

3. Bituminous coal ——————Source of energy.

4. Bauxite ———————————Ore of Aluminum.

5. Shale ————————————Light weight concrete aggregate.

LABORATORY EXERCISE

1. Identify each type of sedimentary rock in your tray using Figure 4-2 and 4-3. Place each sample on a sheet of paper and write the correct name adjacent to the sample.

2. Have your laboratory instructor check your answer for correctness.

3. Write the name of each rock identified on the left margin of a piece of paper and list the possible depositional environments for each to the right of the name. Your instructor may allow you to do this for extra credit.

4. Separate the sedimentary rock samples in your tray into clastic and non-clastic environments.

CHAPTER 5

SEDIMENTARY ROCK FIELD TRIP
by
DOY ZACHRY AND RONALD KONIG

This field trip is intended to provided an opportunity to examine sedimentary rocks in their field occurrences. An introduction to the geology of Northwest Arkansas and to the character of sedimentary units is useful in placing rocks viewed on the trip into their proper geologic setting.

SEDIMENTARY UNITS

The northern Gulf of Mexico today is receiving large volumes of sand, silt and clay delivered from the continent by the Mississippi and other river systems. Such areas, characterized by widespread sediment accumulation, are called sedimentary basins. At various times in the geologic past, sedimentary basins have existed on and adjacent to most of the continental areas of the Earth's surface.

Within a particular basin, sediment accumulates as successive layers on the basin floor. As depositional conditions change, different kinds of sediment are delivered to the basin and deposited as new layers over previously formed ones. The continuation of this process through time produces thick successions of layered or stratified rock such as those seen in the walls of the Grand Canyon or on hillslides surrounding Fayetteville. The geologist uses this characteristic stratification to subdivide thick successions of rock deposited over substantial lengths of geologic time into numerous thinner layers or units that can be studied more conveniently.

The sedimentary succession of Northwest Arkansas is almost 3000 feet thick and rests on a platform of ancient igneous and metamorphic rock. Sedimentation began about 545 million years ago and continued sporadically until approximately 300 million years ago. Sand, silt, clay and carbonate sediment accumulated from time to time as conditions changed and produced a bewildering succession of sandstone, siltstone, shale, limestone, and dolostone layers. The layers of rock have been subdivided into thinner units called formations and members.

A formation is composed of one or more widespread layers that are grouped together for mapping purposes. They are characterized by clearly defined upper and lower boundaries.

A member is a subdivision of a formation although not all formations are subdivided into members. Both formations and members are given formal names after a geographic locality where they are well exposed. Figure 5-1 is a stratigraphic section that depicts the succession of formations and members exposed in Northwest Arkansas near Fayetteville. Remember that each named unit represents a layer or group of layers that are recognized over a substantial geographic area. The boundary surface, usually a bedding plane, that separates one formation from another is referred to as a contact. Thus each formation has two contacts, one at the top and one at the bottom.

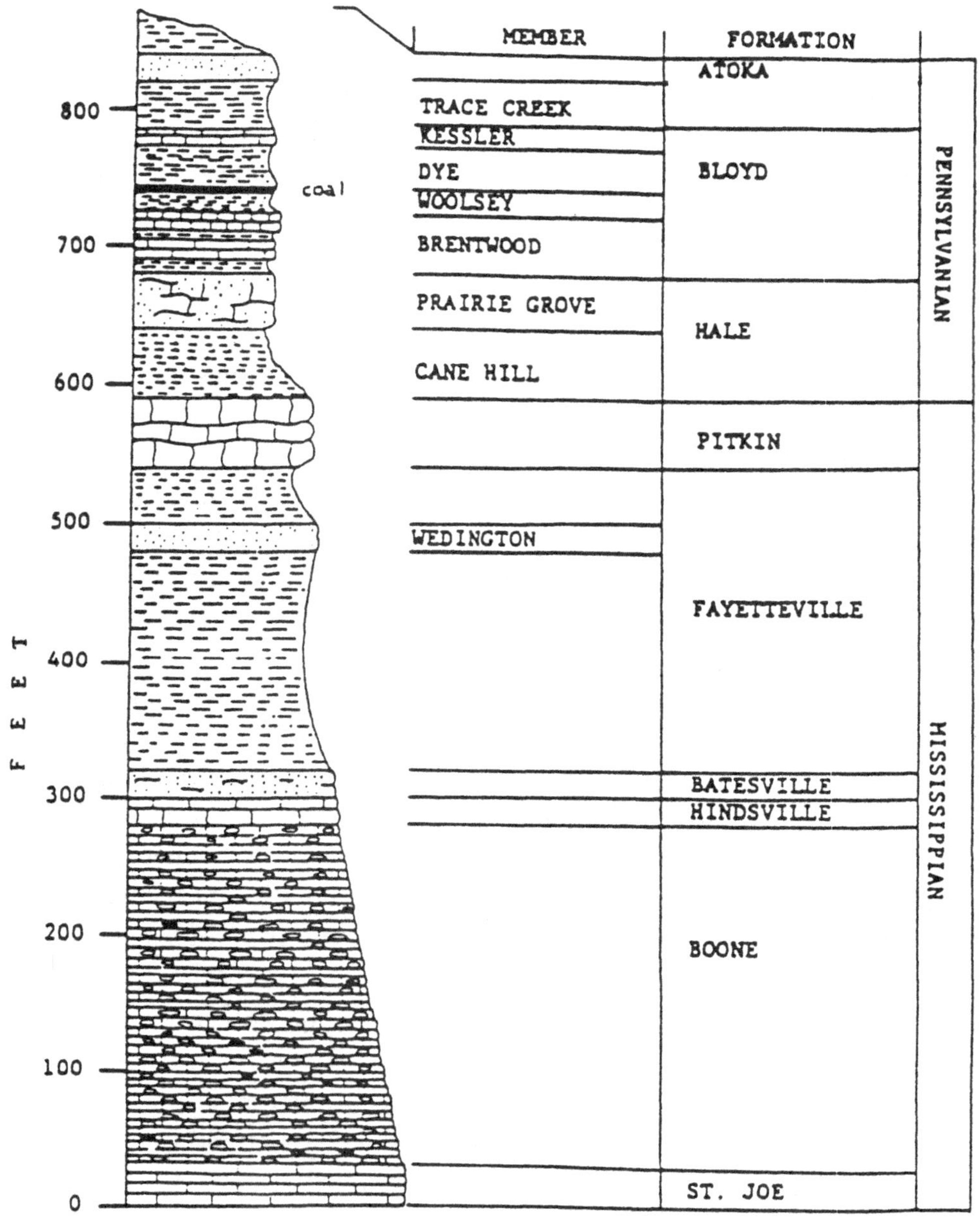

Figure 5-1. Generalized Stratigraphic Section for Northwest Arkansas.

GEOLOGIC TIME

Geologic studies are often designed to reconstruct historical events that have transpired in the geologic past. In order to discuss historical events and to relate them to each other in time, a geological calendar is necessary. The calendar used by geologists for this purpose is based on careful studies of the succession of organisms that have existed in the past, as those organisms are preserved as fossils in sedimentary rock. From the knowledge that certain organisms existed during specific intervals of time, a geologic time scale has been erected. Diagnostic fossils may be collected from rocks of unknown age, and by comparisons with similar fossils known to have lived during particular times, the rocks may be dated and assigned to their position in the historical record.

Using fossil organism as a basis, the last 545 million years of earth history has been divided into large units called Eras. Each Era has been further divided into smaller time units called geologic periods. For example, the Mesozoic Era has been divided into the Triassic, Jurassic and the Cretaceous periods. The geologic time scale is included in Figure 5-2.

Era	Period	Symbol	Millions of Years Ago
Cenozoic	Quaternary	Q	0-1.6
	Tertiary	T	
			66.4
	Cretaceous	K	66-144
Mesozoic	Jurassic	J	144-208
	Triassic	TR	208-245
			245
	Permian	P	245-286
	Pennsylvanian	IP	286-320
	Mississippian	M	320-360
Paleozoic	Devonian	D	360-408
	Silurian	S	408-438
	Ordovician	O	438-505
	Cambrian	Є	505-545
			545
Proterozoic			
Archeozic	Precambrian Time PЄ		

Figure 5-2. Geologic Time Chart showing Eras, Periods, symbols, and radiometrically determined ages.

The formations and members exposed in northwestern Arkansas near Fayetteville accumulated as layers of sediment during the Mississippian and Pennsylvanian periods of the Paleozoic Era. Deep wells drilled near Fayetteville pass through these layers and penetrate older formations of Devonian, Silurian, Ordovician and Cambrian age before encountering the much older Precambrian igneous and metamorphic rocks that lie beneath them. Figure 5-3 is a cross-section that extends from the Missouri border southward to the Arkansas River. This section shows the sedimentary succession of Northwest Arkansas as it is related to the igneous and metamorphic rock below. Note that the Boone Formation underlies the land surface north of Fayetteville. The Northwest Arkansas Mall is constructed on this formation. South of Fayetteville, the Boone Formation is buried by successively younger formations (figure 5-3) and is encountered at great depths in wells drilled near Fort Smith. Sandstone units in the lower part of the Atoka Formation are exposed near the crest of high hills near Fayetteville. Near the Arkansas River, the same sandstone layers are deeply buried and serve as important reservoirs for natural gas.

ROCK SYMBOLS

Standard symbols are used to depict various kinds of sedimentary rock on stratigraphic columns and cross sections. The symbols for several important rock types are indicated below.

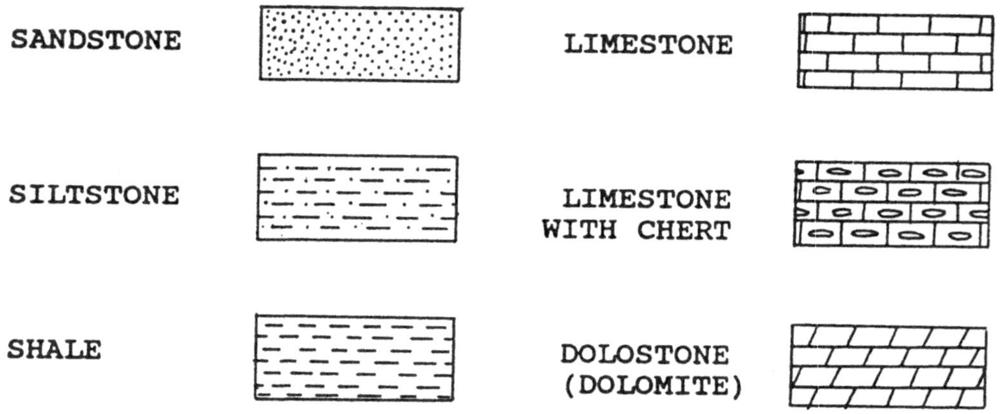

Standard Rock Symbols Used in Chapter 5

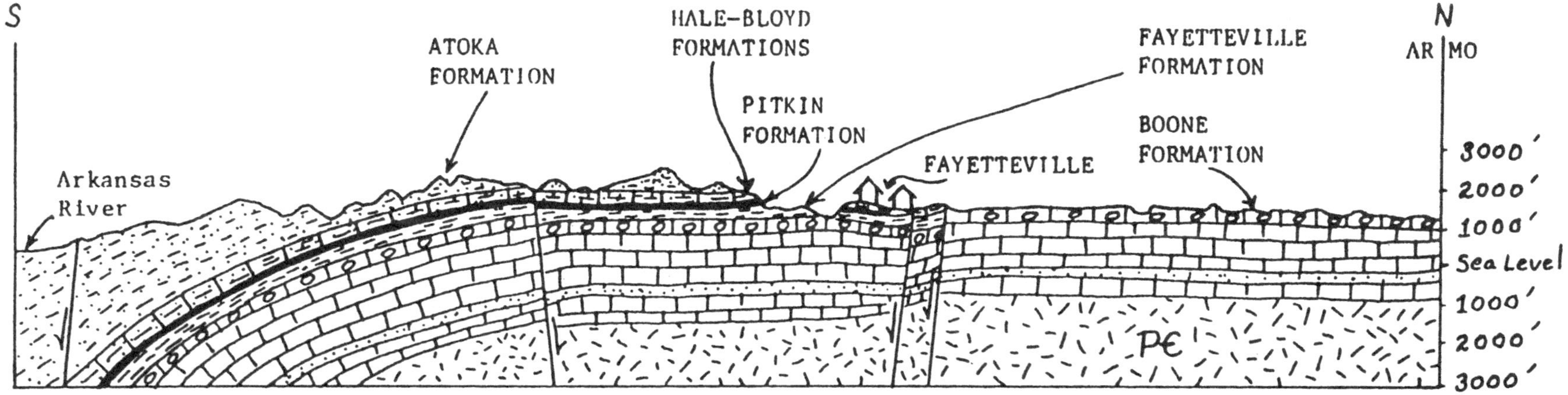

Figure 5-3. Geological cross-section from Missouri border to the Arkansas River. Modified from Cary Croneis, 1930, Geology of Arkansas Paleozoic Area: Arkansas Geological Survey, Bulletin 3.

Railroad Cut, Fayetteville

The railroad cut in Fayetteville begins just north of Maple Street and extends continuously to a point a little north of Dickson Street. A number of sedimentary rock types are exposed here and have been assigned to the Fayetteville, Hale, and Bloyd formations by geologists on the basis of fossil remains and geologic mapping. Locate these rock units on the stratigraphic section (figure 5-1) to determine their relationship to other rocks in the area.

The Mississippian age Fayetteville Formation consists of upper and lower shale members separated by the Wedington Sandstone Member (figure 5-1). At the north end of the railroad cut, the upper member is exposed and consists of gray to black fissile shale that is somewhat silty and contains small claystone and limestone concretions. The Fayetteville Formation is normally overlain by the Pitkin Formation, a gray fossiliferous limestone. At this locality, however, Pitkin limestone has been eroded subaerially and redeposited as a beach gravel consisting of limestone fragments that are now incorporated in the lowermost part of the overlying Hale Formation. Thus the Pitkin Formation is missing at this rock exposure. The contact between the overlying Hale Formation and the underlying Fayetteville Formation is appropriately referred to as an unconformity. The limestone pebbles in this conglomerate have been derived largely from the reworking of the Mississippian age Pitkin Formation, and as such probably represent the northernmost extent of the sea in which sediment was being deposited during Pitkin time. Thus, this conglomerate most likely represents a shoreline depositional environment. The rest of the overlying Cane Hill Member of the Hale Formation is composed of silty shale, siltstone, sandstone and fossil-bearing conglomeratic limestone at the top. (Figure 5-4)

Rocks of the Fayetteville and Hale Formations are inclined gently to the south in this area. Because of this southerly inclination, the rock units exposed along the railroad cut change from north to south. Siltstone and sandstone beds exposed at or close to track level commonly show primary sedimentary structures like stratification, cross bedding, and bottom markings, indicating shallow water deposition and current activity.

Farther south beneath the Lafayette street bridge there is a physical break in the rocks known as the Fayetteville fault. This fault trends northeast and the rocks on its southeast side have been lowered with respect to rocks on the northwest side. The age of this fault is not known, but it must be younger than the Pennsylvanian because it cuts all Pennsylvanian age rock in Northwest Arkansas. There has never been any earthquake activity recorded at seismograph stations that could be attributed to movement along this fault.

Southeast of the Fayetteville fault the younger rock of the Bloyd Formation is exposed. These rocks are inclined toward the north and include limestone of the Brentwood Member, shale of the Woolsey Member (lignite coal does occur at the top of this unit locally), and the Dye Shale Member (figure 5-1).

A generalized north-south cross section has been constructed to aid in understanding the general geology of the area (fig. 5-5). It is assumed in this figure that the railroad tracks are horizontal.

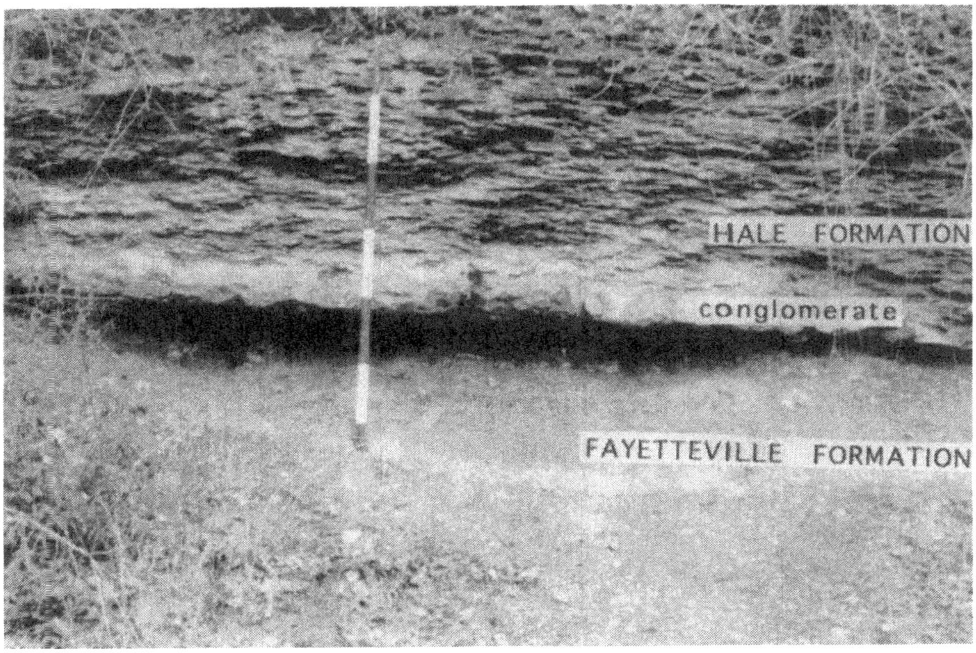

Figure 5-4. Photograph of sedimentary rocks exposed at the north end of the railroad cut. A limestone conglomerate (middle unit), marks the base of the Cane Hill Member. It is about one foot thick, and overlies fissile shale of the upper member of the Fayetteville Formation. Silty shale and siltstone of the Cane Hill Member of the Hale Formation lie above the limestone conglomerate of the Cane Hill Member.

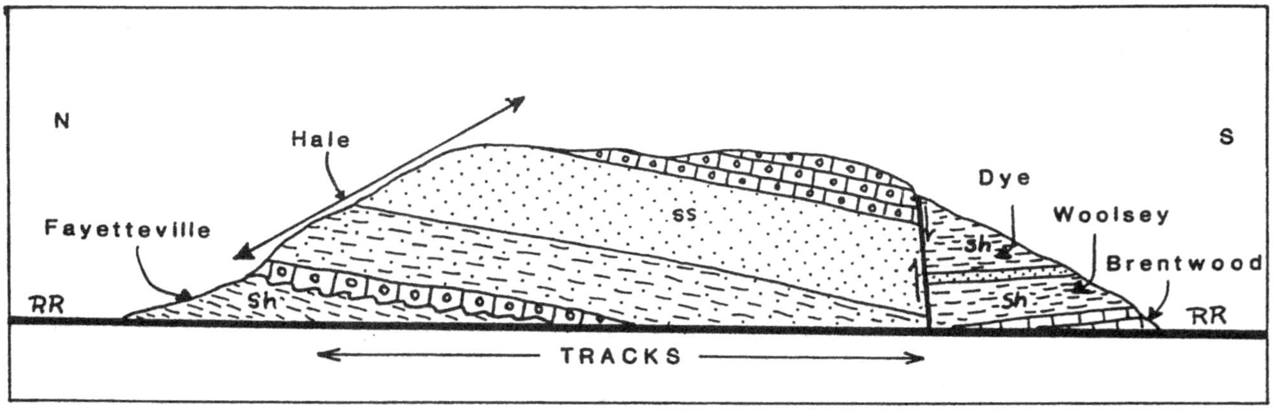

Figure 5-5. A north-south cross sectional diagram showing the general geologic setting in the railroad cut at Fayetteville. The diagram is not drawn to scale.

FIELD TRIP QUESTIONS

NAME ______________________________

Answer the following questions and submit to your laboratory instructor at the beginning of the next laboratory.

1. What kinds of sedimentary rocks are present in the railroad cut? ________________
____

2. What geologic periods are represented ______________________and approximately how old are they in millions of years? ______________________________________.

3. Were these rocks continuously deposited so that there is not a conspicuous time gap in the rock record? ____
____.

4. How thick is the limestone conglomerate that represents the lower part of the Hale Formation? ____.

5. What kinds of sedimentary structures did you observe in the Hale Formation?
____.

6. If we assume the railroad tracks are horizontal , in what general direction are the rocks inclined at the north end of the RR cut ______________________ and how can you tell this in the field?____.

7. What depositional environment existed at the end of the Mississippian Period (circle one)?

Open Sea   Alluvial   Beach   Organic Reef

8. What rock formation and rock type would be encountered next (first) if you drilled a hole downward at the north end of the railroad cut? ____________________________

9. Examine the various types of rock that have been used to provide a base for the railroad tracks between the north end of the railroad cut and the first railroad bridge farther north and answer the following questions:

a) Were all of these rocks derived locally? ______________________.

b) Name at least 6 different types of rock collected at this locality.

1) ____________________ 5) ____________________

2) ____________________ 6) ____________________

3) ____________________ 7) ____________________

4) ____________________ 8) ____________________

10) Identify the building stone used in the following buildings on campus:

a) Entry steps, "Ozark Hall" ______________________

b) Foyer of "Vol-Walker Hall" ______________________

c) Pink rock slabs forming entrance walk to "Bell Engineering Center"

____.

d) Light gray rock forming walls of Home Economics Building

____.

e) Brown rock forming foundation of Old Main______________________.

CHAPTER 6

IDENTIFICATION OF METAMORPHIC ROCKS
by
RONALD H. KONIG

We observed in chapters 3 and 4 how the constituents of rocks are first brought to the Earth's surface as molten material which cools and crystallizes into igneous rock and is then physically and/or chemically broken down into clastic grains (soils) or is put into chemical solution by weathering. Agents of erosion pick up and transport this material to a sedimentary basin where it is deposited as sediment that undergoes lithification to produce sedimentary rock. There remains one basic type or rock to discuss. This rock type forms in nature when any preexisting rock is subjected to changes below the zone of surface weathering. The process by which this change happens is called metamorphism.

As we discuss the actual physical changes within a rock being metamorphosed, it will be useful to keep in mind the idea that a mineral is stable in the environment in which it is formed, and unstable when that environment changes. For a good example of this concept consider a rock composed of the "minerals" sugar, milk, butter, and chocolate. At room temperature this material is stable as a brown mass, but when it is placed in a new environment, an oven, where the temperature is 236 degrees F., the four original constituents reorganize themselves into a new "rock" made entirely of chocolate fudge. The originally stable constituents became unstable when the temperature was raised. Metamorphism usually involves a simultaneous rise in temperature and pressure in the presence of circulating solutions.

The conditions under which metamorphism takes place are not precisely defined because rocks of different composition behave differently under metamorphic conditions. The accepted lower temperature limit for metamorphism is about 150 degrees F. The range of pressures that cause metamorphism is about 1000 bars to about 20,000 bars (one bar is about equal to one atmosphere of pressure). These limits vary with different rock types and also with the amount of water or other fluids within the rock undergoing metamorphism.

The observable property of a metamorphic rock, which tells us that a great directed pressure was applied to it during metamorphism, is foliation. Foliation is the organization of platy minerals, such as mica and chlorite, and tabular minerals like feldspar, into a planner structure so that the rock appears to have layers. In the case of mica, the mineral cleavage would also be parallel to foliation surfaces (figure 6-1). The directed pressure involved in the formation of this foliation should be oriented at an angle of about 90 degrees to the foliation surface.

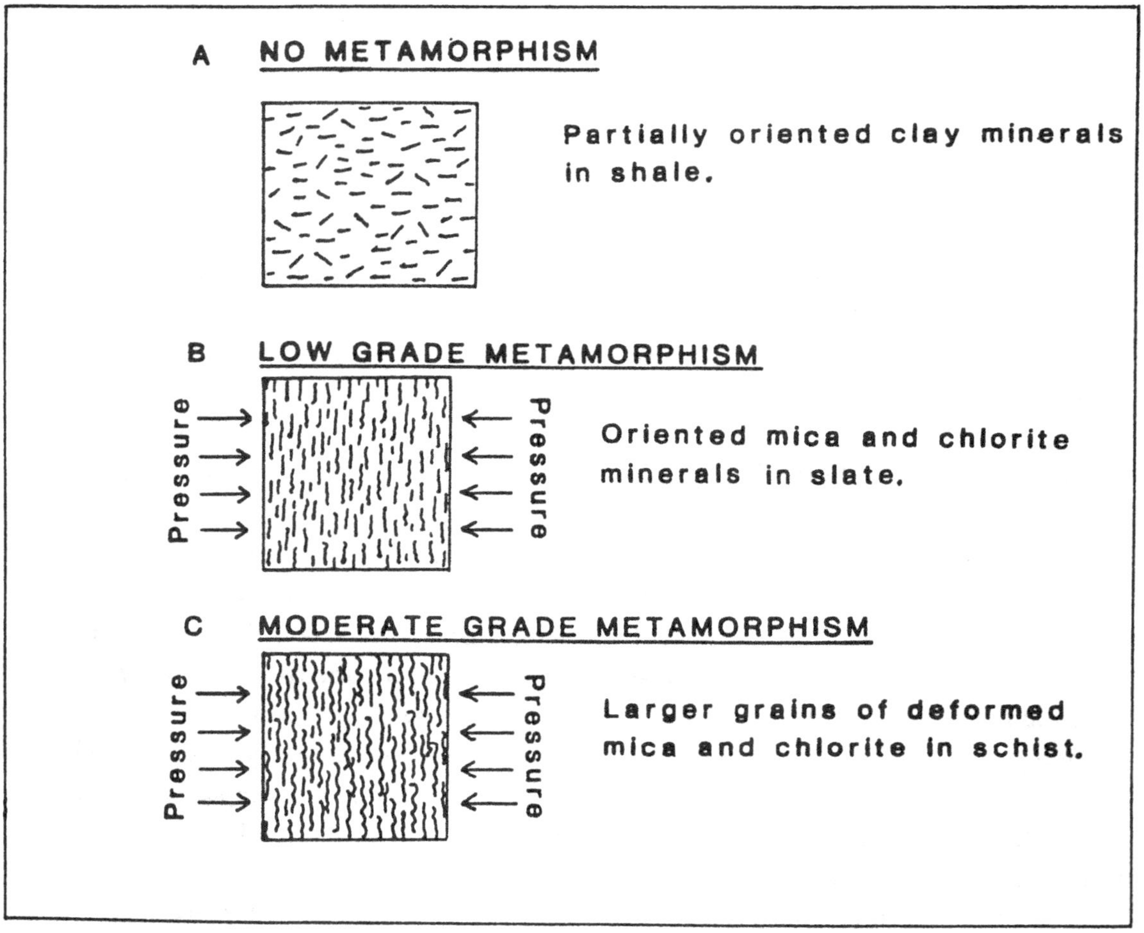

Figure 6-1. Cross sectional sketches showing how foliation results from directed pressure applied to a shale (A) as it changes first into a slate (B) and then into a schist (C) during metamorphism.

TYPES OF METAMORPHISM

Metamorphism most commonly occurs within the Earth's lithosphere where either high temperatures and hot fluids locally invade existing rock, or where a combination of high temperatures and pressures and hot fluids alter existing rock over a wide area. Hence temperature, pressure and fluids are often referred to as the agents of metamorphism. The first, more local type of metamorphism is called contact metamorphism and the second, more widespread type is called regional metamorphism. Each type is discussed below.

CONTACT METAMORPHISM

Contact metamorphism occurs where magma is intruded into existing rock in the earth (figure 6-2). Heat and solutions from the magma slowly invade the country rock and cause it to simply recrystallize if the rock is pure (only one mineral species makes up the original rock), or causes it to recrystallize and form new minerals if the original rock is impure (more than one mineral species makes up the original rock). In both cases the resulting rock will not have any foliation because a uniform or balanced pressure has been important in metamorphosing the existing rock. Typical contact metamorphic reactions may go like this:

A. Layers of pure limestone, when heated by an invading mass of magma will undergo recrystallization. This means that calcite grains in the limestone partially "melt" and grow larger or may simply combine with adjacent calcite grains to form large calcite grains. In either case the resulting metamorphic rock becomes coarser grained, but remains unfoliated. The rock produced is a marble.

B. Layers of shale (contains quartz, clay, and calcite) are heated by an invading magma. Contact metamorphism will cause recrystallization and the formation of a hornfels composed of new minerals such as quartz, feldspar and hornblende as follows:

Clay + Quartz + Calcite ——> Quartz + Feldspar + Hornblende

Naturally different starting materials (rocks) will give different kinds of metamorphic rocks. However, some rock types are more easily metamorphosed than others because they are initially composed of less stable minerals. Minerals like calcite and clay are easily recrystallized or transformed into new metamorphic minerals, whereas quartz and feldspar resist metamorphism. The most common contact metamorphic rocks include marble, dolomitic marble, quartzite (from quartz sandstone) and hornfels (from shale). The first three rock types may also result from regional metamorphism.

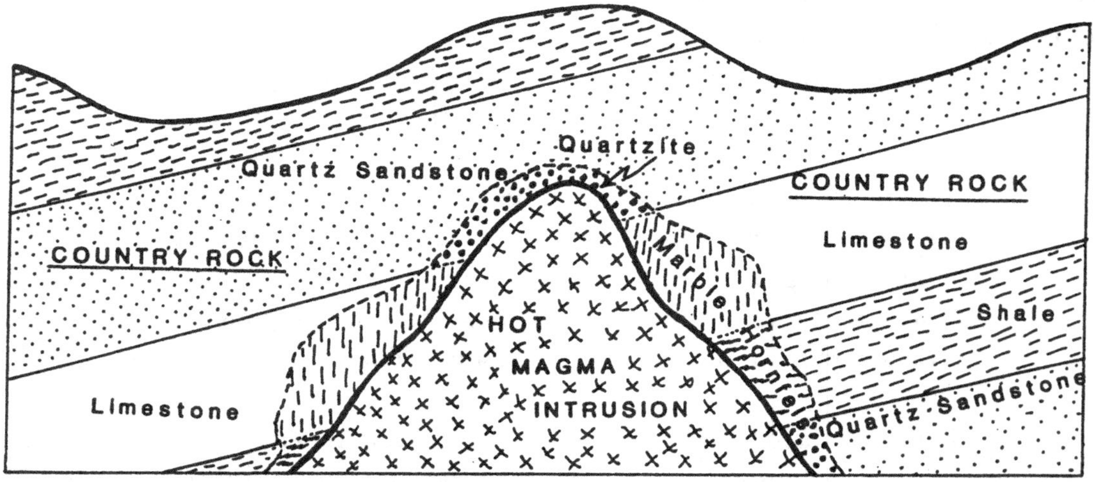

Figure 6-2. Cross sectional diagram showing the zone of contact metamorphism formed by magma in contact with country rock composed of different kinds of sedimentary rock.

REGIONAL METAMORPHISM

This type of metamorphism is regional because the metamorphic effects are seen over a very wide area, whereas contact metamorphism only occurs immediately adjacent to a magma body. Rocks undergoing regional metamorphism are subjected to 1) elevated temperatures because they lie closer to the hotter environment of the Earth's interior, 2) higher pressures due to both the tremendous weight of the rocks on top of them, and the directed pressures associated with converging plates boundaries, and 3) circulating fluids because water is being forced out of water-filled pure spaces and certain water-bearing (hydrous) minerals. Hence the theory of plate tectonics, as described in lecture, goes a long way to explain how and where rocks become regionally metamorphosed (figure 6-3).

Many rocks associated with converging plate boundaries are heated, squeezed and injected with hot fluids to form all of the typical foliated metamorphic rocks. Again the foliation results from directed or unbalanced pressure acting upon existing rock. Platy minerals like mica grow with their "flat" side (cleavage surface) oriented perpendicular to the major regional stress (force), and foliation develops in the rock. As recrystallization and reorientation take place during regional metamorphism, the stable platy minerals grow at the expense of the unstable ones and a preferred orientation becomes increasingly evident in the rock. An example of this process is described below as a shale undergoes progressively stronger regional metamorphism.

A shale composed mostly of kaolinite (clay) with some feldspar and quartz silt is involved in plate collision. As pressure and temperature build up, small clay minerals of the shale recrystallize into slightly larger mica grains and orient themselves in a stable position. The shale has now been metamorphosed to a slate and a foliation (rock cleavage) is observable. With increased pressure and temperature the oriented mica continues to recrystallize and grow larger as metamorphism proceeds until they become just visible to the unaided eye. At this point a noticeable preferred orientation has developed and the rock is now called a phyllite. As the grade of metamorphism increases further, the mica flakes continue to recrystallize and grow quite large, and any quartz, feldspar, or other mineral that was originally present is also recrystallizing and growing larger. A strong foliation is now observable and the rock is called a schist. If metamorphism still continues, the quartz and other non-platy minerals will begin to separate from the mica and produce definite bands of different minerals. A rock showing this mineral banding is called a gneiss. This change from shale to slate to phyllite to schist to gneiss is an excellent example of progressive metamorphism. This and other examples of progressive metamorphism are illustrated in figure 6-4.

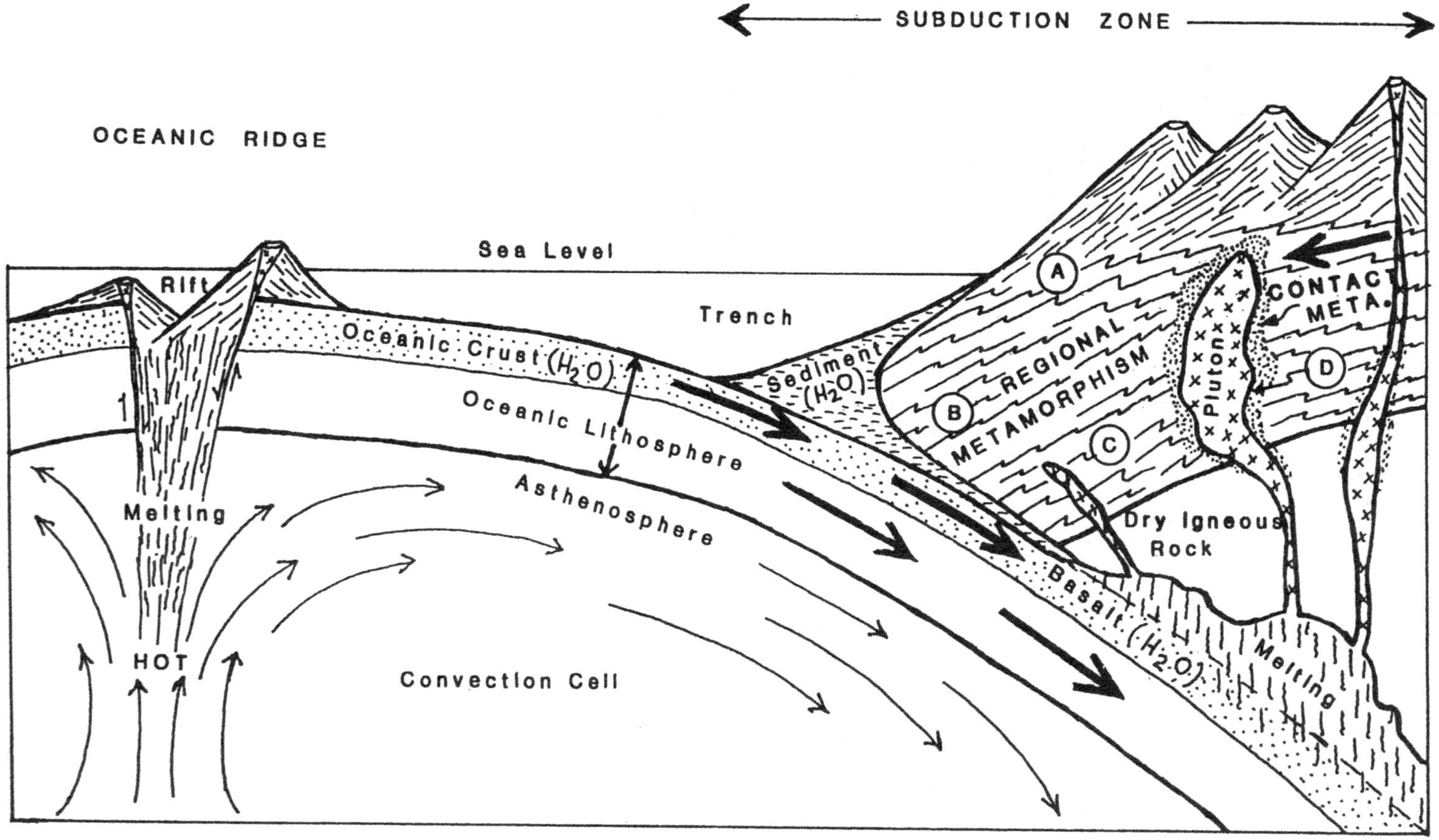

Figure 6-3. Cross sectional diagram showing a spreading boundary (oceanic ridge area) and converging boundary (subduction zone). Regional metamorphism occurs at A (low temperatures and pressures), B (moderate temperatures and pressures), and C (high temperatures and pressures). Contact metamorphism occurs at D (stippled areas).

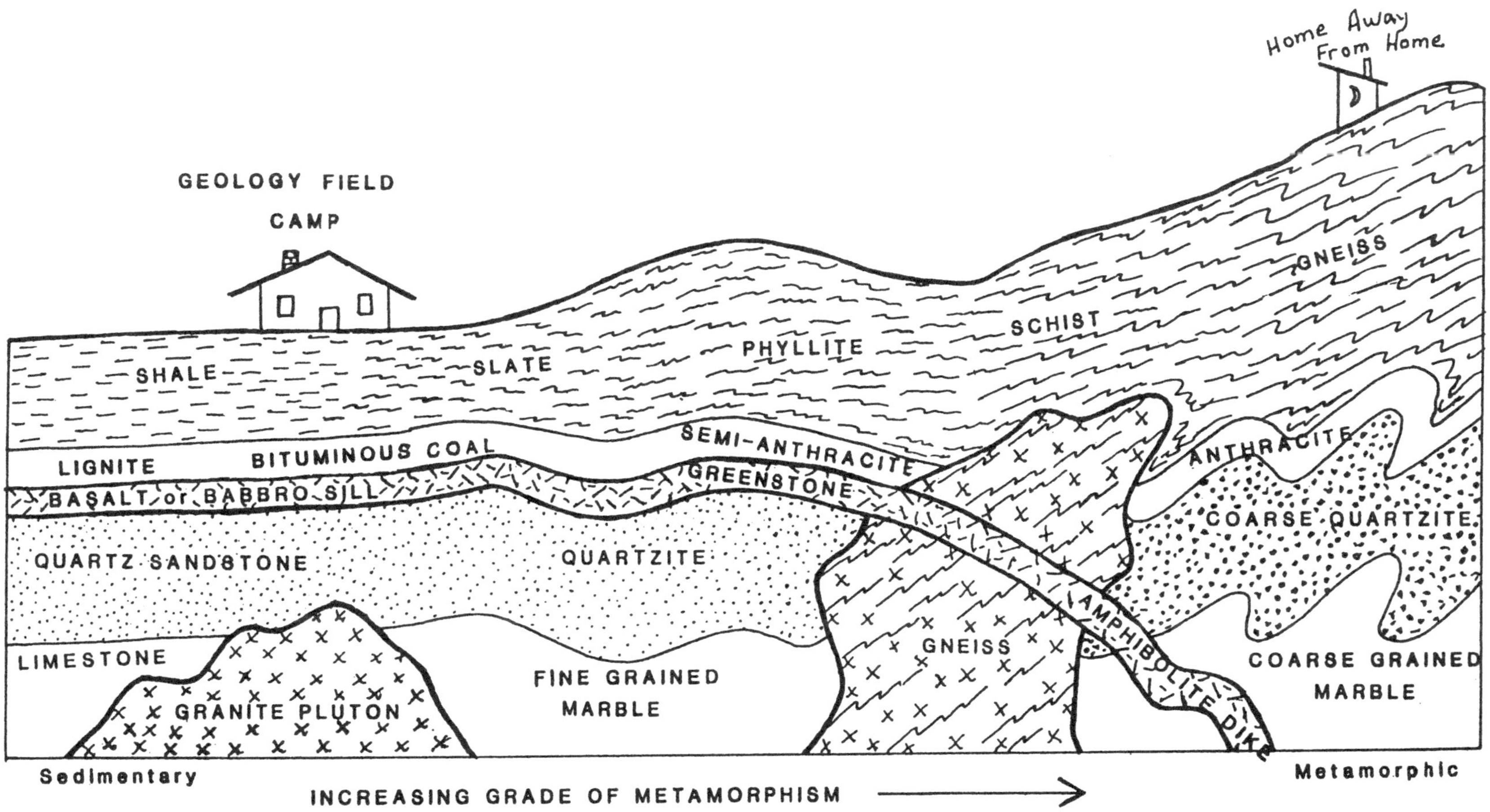

Figure 6-4. Cross sectional diagram showing what happens to shale, lignite, basalt or gabbro, quartz sandstone, limestone and granite when subjected to progressive regional metamorphism.

The rock types named in the above metamorphic synthesis are important to know because they will aid you in lab as you construct a hypothesis about the origin of a regionally metamorphosed rock. Figure 6-5 illustrates the major temperature and pressure environments related to regional metamorphism of a shale.

In viewing figure 6-5, note that minerals recrystallize at higher temperatures the farther the rock lies beneath the surface of the earth. This relationship exists because increasing pressure on the rock makes its minerals more stable with respect to recrystallization. The same relationship exists when dry igneous rock is melted under high pressure.

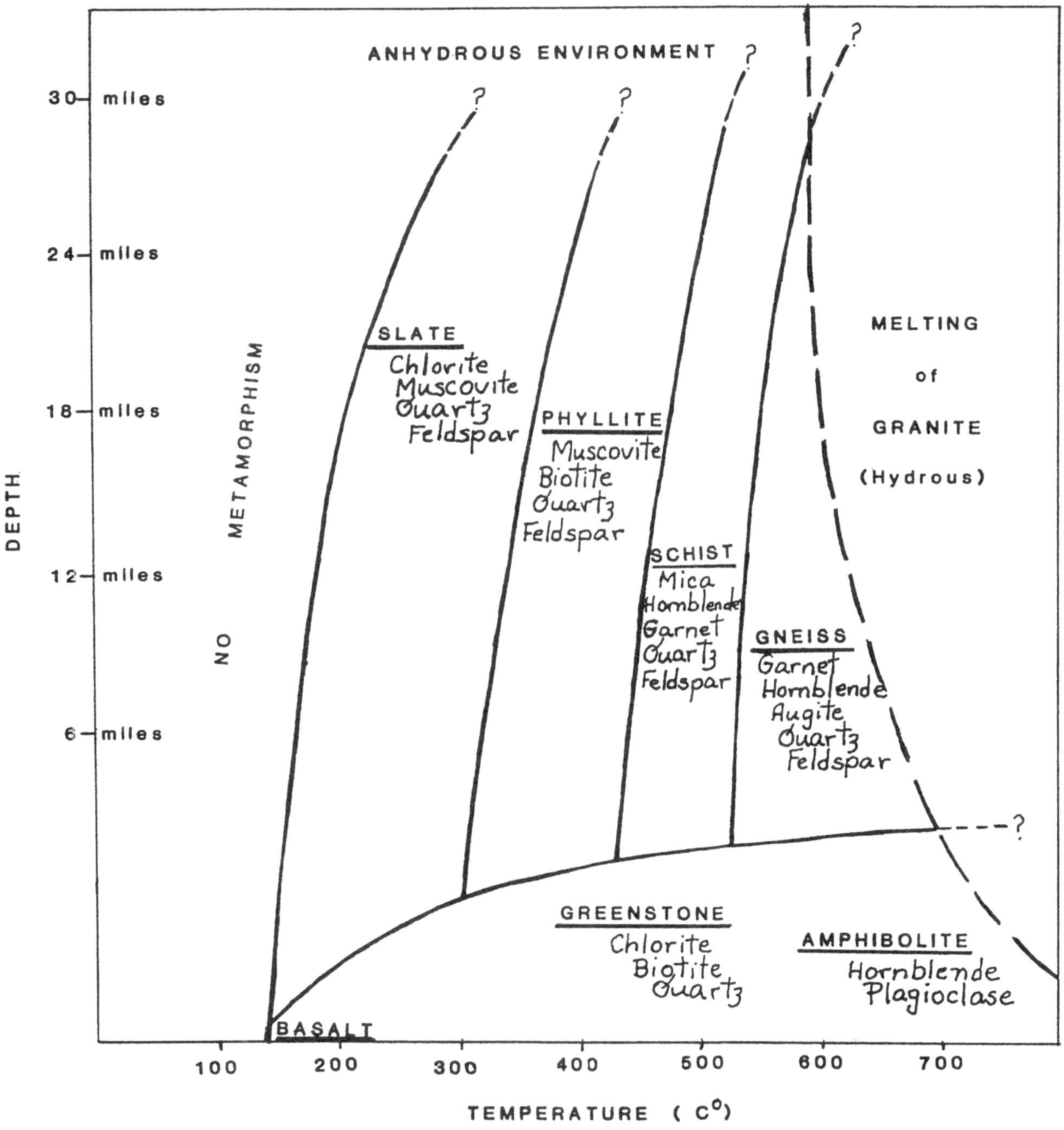

Figure 6-5. Generalized diagram showing the relationship between temperature and pressure during regional metamorphism of a shale and basalt or gabbro. Stable minerals are listed under each metamorphic rock type.

CLASSIFICATION

Metamorphic rocks are divided into two major groups based on the presence or absence of foliation. Texture (basically grain size), structure (basically the appearance of foliation), and mineral composition are then evaluated to finally determine the name of the rock. Figure 6-6 is a convenient identification chart to be used in classifying all metamorphic rocks in this laboratory exercise.

TEXTURE and STRUCTURE		VISIBLE MINERALS	ROCK NAME
STRONGLY FOLIATED	VERY FINE GRAINED (cleavage)		SLATE
	FINE GRAINED (cleavage)		PHYLLITE
	COARSE GRAINED (wavey layers)		SCHIST
	COARSE GRAINED AND BANDED	GARNET, AUGITE, HORNBLENDE, FELDSPAR, MICA, QUARTZ	GNEISS
WEAK TO MOD FOLIATION	FINE GRAINED	CHLORITE and FELDSPAR	GREENSTONE
	FINE TO MEDIUM GRAINED	SERPENTINE	SERPENTINITE
	MEDIUM TO COARSE GRAINED	HORNBLENDE, QUARTZ and FELDSPAR	AMPHIBOLITE
NON FOLIATED	GLASSY EXTREMELY FINE GRAIN.	NONE	ANTHRACITE COAL
	FINE GRAINED	QUARTZ and MICA	HORNFELS
	FINE TO COARSE GRAINED	CALCITE or DOLOMITE	MARBLE or DOLOMITIC MARBLE
	MEDIUM TO COARSE GRAINED	QUARTZ	QUARTZITE

Figure 6-6. Identification chart based on mineral composition and texture/structure for most common metamorphic rocks. Rocks with an * are not in your rock and mineral tray.

DESCRIPTION OF METAMORPHIC ROCK TYPES

Slate	Very fine-grained rock with well-developed rock cleavage that may cut bedding at any angle. Individual minerals may not be identified without a microscope. Color of rock is typically gray, but may be red, green, or tan.
Phyllite	Fine-grained rock with rock cleavage similar to that of slate, but usually showing very small crinkles (folds) on cleavage surface. Tiny grains of muscovite mica often seen on cleavage surfaces in reflected light. Color of rock is typically gray.
Schist	Coarse-grained rock with wavy (folded) surfaces called schistosity. Mica and/or chlorite (very abundant), quartz, garnet, feldspar, and hornblende can usually be recognized. Color is usually some shade of gray.
Gneiss	Coarse-grained rock with conspicuous light and dark colored bands. These bands are composed of quartz and feldspar (light bands) and hornblende, augite and/or biotite mica (dark bands).
Hornfels	Rock is very fine grained, lacks foliation, and is composed of tiny grains of mica and quartz when derived from the contact metamorphism of shale. Dark spots that stand out in relief are common. Rock may otherwise resemble a basalt.
Anthracite	Shiny rock composed of carbon and produced by metamorphism of bituminous coal. Rock is dark gray to black, will not leave a mark on white paper, has a hardness of about 3, and is often confused with obsidian.
Marble	Fine to coarse-grained rock composed of interlocking grains of calcite. Rock is usually white to light gray and will always react vigorously with dilute hydrochloric acid. May have a few small grains of chlorite.
Dolomitic Marble	Rock is similar to a marble except it is composed largely of dolomite. Dolomite will give a strong reaction to dilute hydrochloric acid when a small portion of the rock surface is powdered with the point of a nail.
Quartzite	Rock is medium to coarse-grained and has resulted from the metamorphism of a quartz sandstone. The interlocking quartz grains make this rock tough and very expensive to drill and excavate. Quartzite is often light gray, but may be tan, brown, red, or green. Regardless of color, fractures pass through individual quartz grains and not around them.

Greenstone Rock is fine grained, weakly foliated and some shade of green. Chlorite and feldspar are present, but due to grain size are not recognizable. Low to moderate regional metamorphism of basalt and gabbro forms a greenstone.

Amphibolite Rock is medium to coarse-grained, weak to moderately foliated and medium gray to black. Quartz, plagioclase feldspar and hornblende dominate this rock. Moderate to high grade regional metamorphism of basalt and gabbro forms amphibolite.

Serpentinite Dark green rock composed mostly of the mineral serpentine, that has been derived from the metamorphism of olivine and/or pyroxene in basalt, gabbro, peridotite, or dunite.

USES OF METAMORPHIC ROCKS

1. Slate————Roofing shingle, walk paving, billiard tables
2. Anthracite———Fuel
3. Marble————Dimension stone, monuments
4. Quartzite————Crushed stone, concrete aggregate
5. Serpentinite———Verde Antique (green marble)

LABORATORY PROCEDURE

1. Remove the metamorphic rocks from the rock and mineral tray and place them on the table.

2. Separate them into three groups known as
   a) strongly foliated
   b) weakly to moderately foliated
   c) non-foliated

3. Place the group of rocks having strong foliation on a piece of plain paper and:
   a) identify conspicuous minerals
   b) establish grain size for each sample
   c) Name each rock by referring to figure 6-6.

4. Repeat step 3 for the remaining weakly to moderately foliated and non-foliated rocks.

5. A thin section of a metamorphic rock is mounted on the stage of petrographic microscope in the laboratory. Focus the microscope for your eyes and then draw a diagram of a portion of the viewing area that adequately shows grain relationships of this rock.

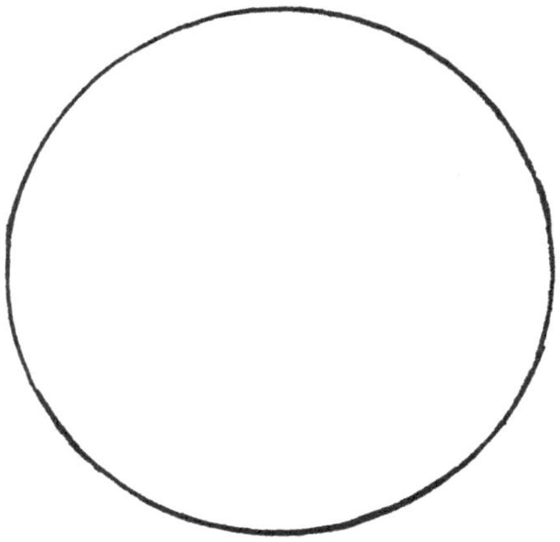

6. Assume that all of your rocks having foliation formed at a depth about 12 miles below the Earth's surface. Determine the maximum and minimum temperature necessary for the formation of each of the foliated rocks (slate, phyllite, schist, and gneiss) by referring to figure 6-5. Write this temperature range under each of the rock names determined in question number 3c above.

7. Determine the original rock type (parent rock) from which each dark colored metamorphic rock sample was derived by examining figure 6-4. Write this name below the temperature of formation for the sample.

8. Determine the most likely place of origin for <u>six</u> (6) of your rock samples by examining figure 6-3. The letters A, B, C, and D in figure 6-3 refer to areas of metamorphism. Place the appropriate letter for each rock sample below the parent rock name determined in questions 3 and 4 above.

9. Have your laboratory instructor check your work before leaving.

10. If you have extra time it is advisable for you to review rocks and minerals covered in previous weeks to prepare for your rock and mineral final exam next week.

CHAPTER 7

REMOTE SENSING

by

RONALD H. KONIG AND ALEXANDER TAYLOR

INTRODUCTION

Remote Sensing, the science of collecting and interpreting information about an object, without ever coming in contact with that object, has become increasingly important to humans during the last 50 years. The reason for this is related to our increasing need for knowledge about such things as 1) topographic features (mountains, hills, valleys, etc.), 2) hydrologic features (rivers, streams, lakes, etc.), 3) cultural features (roads, buildings, dams, etc.), and 4) various features of military significance. Much information about any of the above features may be obtained by simply studying images gathered earlier by airplane or satellite.

HISTORY OF REMOTE SENSING

The science of Remote Sensing had to wait for the development of the first practical camera by Francois Deguerre in 1837. Initially this camera was used primarily for the purpose of entertainment. However during the American Civil War, photographers in observation balloons used this camera to great advantage by accurately recording the location of enemy troops. It was not until World War I, however, that extensive use was made of black and white aerial photographs (remote sensing images). This technique was so successful in World War I that more modern cameras loaded with both black and white and color film, were used to locate and plot the movement of troops during World War II. As techniques improved, additional types of photographs (images) were developed. First came color infrared imagery that is useful in distinguishing between objects of the same color but different composition. About the same time thermal infrared imagery, which records levels of heat emitted from objects was developed. Finally, imaging radar was perfected and put into general use in the 1960s. Radar images are particularly useful because radar "sees through" haze, clouds, smoke, snow and even dry soil. Thus areas normally obscured by clouds, etc. can be imaged and studied.

ADVANTAGES OF REMOTE SENSING IMAGERY

There are several advantages associated with remote sensing. A few of the advantages are listed below.

1. It allows us to see topographic highs and lows through the effect of stereo (3-D) vision.
2. It gives us a good vantage point and a greater field of view.
3. It records data permanently so that changes can be noted over longer periods of time.
4. It allows us to view remote locations in a short period of time at a relatively low cost.

USES OF REMOTE SENSING IMAGERY

Because remote sensing images range from standard black and white photography to more complex radar imagery, they have hundreds of useful applications. Some of the more common uses of remotely sensed images include the following:

1. Mapmaking
2. Location of building sites
3. Damage assessment associated with water, wind, lava, gravity, earthquakes, etc.
4. Crop surveys (production estimates and disease extent)
5. Recognition of vegetation types and distribution of disease
6. Military reconnaissance
7. Geological Investigations
    a) Mapping of different rock types and structures.
    b) Development of Geologic models for:
        1) Petroleum exploration
        2) Mineral exploration

SCALE OF REMOTELY SENSED IMAGES

Some of our lab work this week involves imagery taken from space (satellite imagery) and some with photographs taken from airplanes. Airplane images may be high altitude (40,000 feet plus) or low altitude (less than 40,000 feet). Clearly the scale for each of the above images is different. Satellite images are very small scale, whereas, low altitude images are large scale representations of the Earth's surface. This means, that a certain length line on the image equals a certain length line on the ground surface. If the observer knows the actual distance between any two points on the ground, the scale of the image can be determined. For example, if the distance between two houses along a road measures one inch on the image, and the actual ground distance between these houses is known to be one mile, then the scale of the image is one inch equals one mile. On a large scale image one inch on the image would equal, say, one mile on the ground surface, whereas on the small scale, image one inch on the image would equal, say, five miles on the ground surface. The ratio between the image distance and the ground surface distance is known as the fractional scale of the image and can always be determined. This scale is usually referred to as the representative fraction or R. F. on any image or map.

SAMPLE PROBLEM: It is determined that 3.5 inches on an image or map equals 1.5 miles on the ground surface. We need to determine the R. F. of the image or map.

Solution:

1) We know that:

$$\text{R.F.} = \frac{\text{image distance}}{\text{ground distance}} \quad \text{or} \quad \frac{\text{3.5 inches}}{\text{1.5 miles}}$$

2) Conversion to similar units. Because it is not correct to divide inches by miles, it is necessary to change the 1.5 miles of the denominator to inches. There are 63,360 inches in one mile.

Therefore

$$1.5 \text{ X } 63{,}360 = 95{,}040 \text{ inches}$$

3) Thus:

$$\text{R.F.} = \frac{\text{image distance}}{\text{ground distance}} \quad \text{or} \quad \frac{\text{3.5 inches}}{\text{95,040 inches}}$$

4) Reduction of numerator to unity. In equation (3) the numerator is always expressed as unity or one. Thus it is necessary to divide equation (3) by 3.5.

Thus:

$$\frac{\text{3.5 inches}}{95{,}040} \div \frac{3.5}{3.5} = \frac{1}{27{,}154}$$

5) Therefore:

$$\text{R.F.} = \frac{1 \quad \text{(units on image)}}{27{,}154 \quad \text{(units on ground)}}$$

STEREOSCOPIC VIEWING

The value of remotely sensed images is enhanced when two images (called a stereo pair) are used in conjunction with a stereoscope to obtain a three dimensional view of the Earth's surface. Aircraft involved in obtaining imagery fly along parallel flight lines taking images of the land surface. Consecutive images along the flight line are taken so that each image overlaps the next image by 60 percent (forward lap). Figure 7-1 shows the nature of this forward lap. The overlap area on two consecutive images will produce three dimensional observation when viewed with a stereoscope. Some skilled observers are able to see 3-D images without the aid of a stereoscope because they have learned to keep lines of sight from each eye parallel and still bring images into focus.

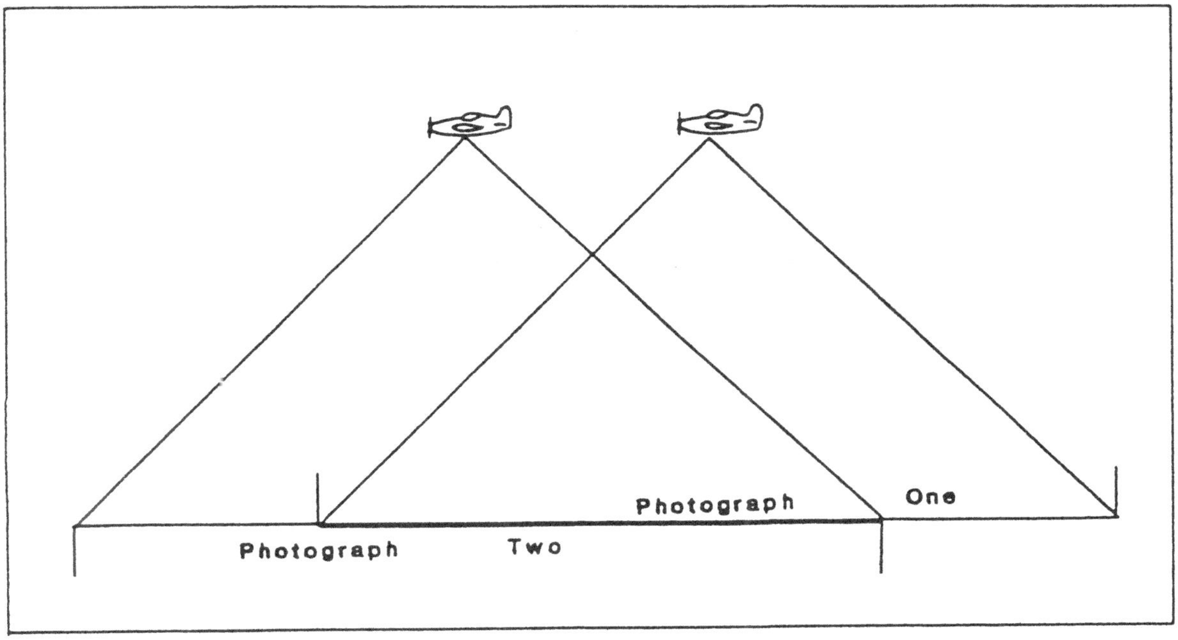

Figure 7-1. Aircraft making overlapping images for the purpose of producing stereoscopic viewing. Three dimensional observation occurs in overlap area of the two images.

Images in adjacent flight lines have about 30 percent overlap. This overlap (sidelap) is generated to assure that all objects along adjacent flight lines have been covered. It has nothing to do with three dimensional observation.

When stereo pairs with 60 percent forward lapping ground coverage are examined with a stereoscope in such a way that one eye sees objects on only one photograph and the other eye sees the same objects on the other photograph, the brain combines both images to produce a three dimensional image. Hills, valleys, buildings, and other features stand out in sharp relief. All of these features appear exaggerated to the observer (trees and buildings appear too tall and valleys appear too deep) because both the photographic process and the magnifying stereoscope used for observation generate vertical exaggeration. For example, the small (pocket) stereoscope magnifies all objects on the photograph about 4X, whereas, the larger mirror stereoscope magnifies objects only about 2X.

LABORATORY EXERCISE

1) A number of forward lapping images are distributed around the laboratory. Each stereo pair has been fastened to the table in such a way that observation of the stereo pair with a stereoscope should provide the viewer with a three dimensional image. If either you or your partner do not see a very conspicuous three dimensional image, ask your laboratory instructor for further explanation. If 3-D viewing is possible, answer the questions associated with each pair of photographs on a separate sheet of paper.

2) Additional images are fastened to the wall. These images do not produce three dimensional viewing because there is no overlap of adjacent frames. Each student team should answer the questions associated with these images on a separate sheet of paper.

CHAPTER 8

TOPOGRAPHIC MAPS

by

RONALD H. KONIG AND ROY B. VANARSDALE

INTRODUCTION

A map is a two dimensional representation of a part of the Earth's surface on which the shape, size, and relative position of natural and cultural features are illustrated. Maps are usually designed for different purposes. Highway maps emphasize highway routes and provide less detail for many natural features. Drainage maps emphasize streams and rivers with less emphasis placed on highways and cities.

Geologists and engineers involved with geologic mapping projects and construction projects, such as highways and dams, are especially concerned with maps that show the configuration of land surface. Such maps are called topographic maps. These maps show cultural features, such as highways, railroads, and cities, as well as natural features, such as streams and rivers. In addition, topographic maps depict, by means of contour lines, the size and shape of hills, mountains, and stream valleys. They also show either the actual or approximate elevation above mean sea level of any geographic point.

Topographic maps are valuable tools for geologists in that the configuration of various landforms may provide clues about the erosional and depositional processes that were important in their formation. They also provided clues about the kind of rock and the geologic structures that lie beneath the Earth's surface. Hydrologists use topographic maps to study streams and to determine stream gradients. Engineers planning highways use topographic maps to select the least costly routes between various points in that they are able to avoid high hills and extremely steep slopes.

Topographic maps also are of great value from a recreational stand point. Hunters, fisherman, hikers, and canoeists use them to examine the topography of areas they frequent.

The most common and widely used topographic maps in the United States today are prepared by the United States Geological Survey (U.S.G.S.). They are called topographic quadrangle maps and are constructed so that their northern and southern boundaries are parallel to lines of latitude and their eastern and western boundaries are parallel to lines of longitude. Standard symbols and scales are used on these maps to depict various features.

Topographic maps generated by the United States Geological Survey are available for most parts of the country. Index maps for each state showing the mapped areas are also available, and may be obtained free from:

U.S. Geological Survey
Map Distribution
Federal Center, Building 41
Denver, CO 80225

From these state index maps, individuals can select specific maps that cover their area of interest and order them from the Survey office. As an option, the user may purchase these maps either from local sporting goods stores or a State Geology Survey Office at a cost of about $4.00 (current price, 1997)

SYMBOLS ON TOPOGRAPHIC MAPS

The use of topographic maps for either geological or recreational purposes requires an understanding of the various colors, symbols, and scales that relate to various features on the map. Some of the more important items include:

1. Colors. Six standard colors of ink are used on most U.S.G.S. topographic maps. Cultural features are shown in black. They include houses, churches, and schools, as well as political boundaries. Water features, such as streams and lakes are in blue. Elevation above mean sea level and contour lines are shown in brown. Green is used to depict vegetated areas. Major highways are shown in red and densely populated city areas are shown in pink on more recent maps. For a more complete understanding of the use of the colors on topographic maps, the student is referred to the Topographic Map Symbols Chart on the following page.

2. Symbols. Numerous cultural and natural features are shown on U.S.G.S. topographic maps by symbols. The chart on the following page shows the standard symbols used to depict most features. Commonly occurring ones such as for houses, churches, schools, highways, railroads, and cemeteries should be carefully noted and remembered.

3. Scales. U.S.G.S. Topographic maps are drawn accurately to scale. Each linear distance on the ground is represented by a much smaller distance on the map. The scale of each map is shown by a "bar scale" or graphic scale as well as by a fractional scale, better known as the representative fraction or R.F. The bar scale is composed of several lines that are subdivided into feet, miles, and kilometers. Distance between points on the map can be determined by comparing them directly with these "bars". The representative fraction simply indicates the ratio between distances on the ground and distances on the map. The fraction 1:24,000

means that 1 inch on the map represents 24,000 inches on the ground, or one centimeter on the map represents 24,000 centimeters on the ground. Some maps make use of a third type of scale, known as the verbal scale, where the scale is stated as one inch equals two miles, etc. on the map. U.S.G.S. topographic maps are prepared in several scales. The most common scale used today is 1:24,000. The map is called a 7.5 minute quadrangle map because it is 7.5 minutes of latitude in length and 7.5 minutes of longitude in width. A quadrangle may, however, be 15 or 30 minutes long and wide. One degree and 2 degree maps are also available.

To determine distances between two geologic points, simply determine this distance on the map and compare it directly with the "bar scale". If one is concerned with distances that can not be measured in a straight line, such as the length of a stream, a piece of string may be used. Place the string on the map in such a way that it directly overlies the curving course of the stream. Then compare the total length of string used with the "bar scale" to determine the length of the stream in feet, miles, kilometers, etc.

LOCATION OF POINTS ON MAPS

The location of particular points or areas and the verbal expression of that location is important for the purpose of defining property lines for general and legal reasons. Various grid systems are placed on maps for this purpose. The most detailed system commonly used in the United States and inaugurated by Congress in 1784, is the Public Land Survey System. This system is used in all but the original 13 states and in a few southwestern states. The essential definition of this system is that land is divided by north-south lines drawn parallel to a principle meridian, and by a second set of east-west lines drawn parallel to a base line that crosses the north-south lines at right angles. These intersecting lines form rectangles that are approximately six miles on a side. A tier of rectangles extending east-west is called a township, and a tier of rectangles extending north-south is called a range. Each rectangle is numbered with respect to a nearby principle meridian and base line (figure 8-1A). The location of the six mile square in which the "X" is placed in figure 8-1A would be T2N, R3W. Each rectangle (TOWNSHIP) is further subdivided into 36 sections, each one mile on a side and numbered as indicated in figure 8-1B. Subdivisions of sections are made by describing the position of a point or area in the section such as the southeast quarter of section 8 (figure 8-1C). This is written NE 1/4, sec 8. Each section (a square mile) is 640 acres, each half-section is 320 acres, and a quarter section is 160 acres.

Townships and ranges are designated on topographic maps. Each of these townships is subdivided into sections and the sections are usually numbered. Further subdivisions are not present on the maps and must be estimated by using imaginary lines that subdivide the section.

Thus the "X" of figure 8-1C Is designated as being in the SW 1/4, SE 1/4, SEC. 8, T.2N., R.3W.. This restricts a point or area to a 40 acre block in terms of location. Carefully study figure 8-1 to gain familiarity with this system, and see how the 40 acre block is located by this system of subdivision.

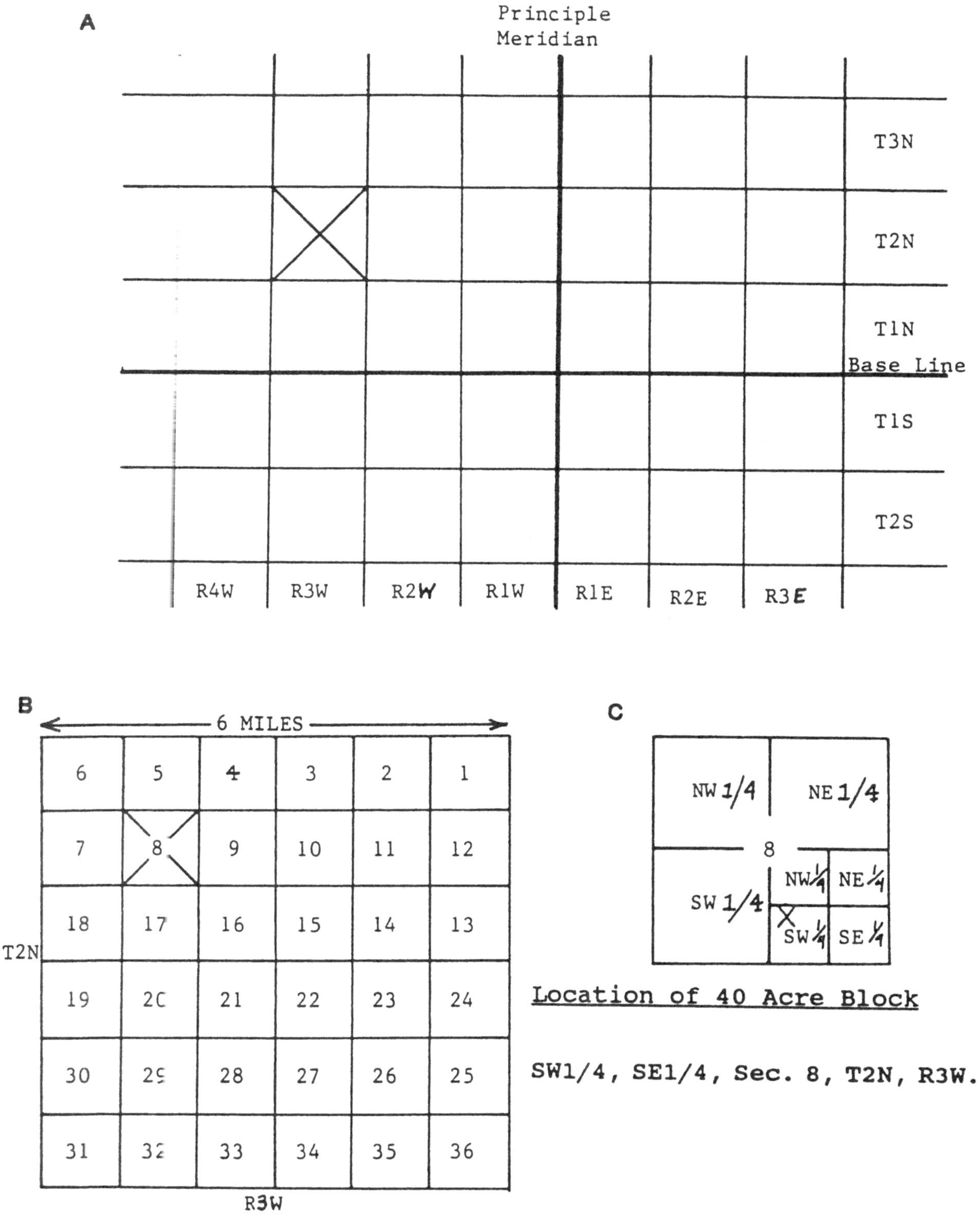

Figure 8-1. Public Land Survey Systems showing townships, ranges, sections, and quarter sections.

ELEVATIONS AND CONTOUR LINES

Elevation is defined as the vertical distance above an established datum. The datum for U.S.G.S. topographic maps is mean sea level. As areas are surveyed, permanent points are established and their elevation precisely determined. These points are called bench marks and are indicated along with their elevation on topographic maps. Elevations are also determined for various points that are not marked with permanent fixtures. These points of "spot elevations" are located on maps by a symbol and their elevation values indicated. Consult the Topographic Map Symbols Chart for the symbols used to illustrate bench marks and spot elevations. Contour lines, bench marks, and spot elevations provide the elevation controls necessary for the use of a topographic map.

Relief is a convenient term used to express the elevation difference in a particular region. It is determined by finding the lowest and highest points in the area and determining the elevation difference between them.

The configuration of the land surface is represented on topographic maps by contour lines. An understanding of these lines and their meanings is critical to topographic map reading. A contour line may be defined as an imaginary line on the ground, every part of which is at the same elevation above sea level. The shore line of any relatively stable body of water is in effect a contour line. A contour line could be drawn at any elevation, but in practice, only the contours at certain regular intervals of elevation are shown. The difference in elevation or vertical distance between adjacent contour lines is called the contour interval. Contour intervals of 10, 20, 50, and 100 feet are used on U.S.G.S. topographic maps. The interval for a specific map is indicated at the bottom near the scale and is designated "Contour Interval 20 Feet". Before examining a particular map, the contour interval for that map should be noted as it may be different from maps studied previously.

To make contour lines more easily used, every fifth line is printed heavier than the others and is labeled with an elevation number. This line is called an index contour. The lighter lines are called intermediate contours and are not labeled.

Each contour line has a specific elevation value, and all points on that line are at the same elevation above mean sea level. If one walks on the ground along a contour line on the map, one is always at the same elevation. If one crosses contour lines, one is either moving uphill or downhill. Contour lines on gentle slopes are widely spaced, whereas, contour lines on steep slopes are very close together. Figures 8-2 and 8-3 show a hill as depicted by contour lines and as seen in profile or side view. Note that the steep slope is reflected by closely space lines. Note also that the value of any contour line can be determined by counting up or down from an index contour line.

The elevation of any point on a contour line can be established by finding the elevation value of that line. The elevation of points falling between contour lines must be reported as ranging between the elevation values of lines directly above and below the point. For example,

TOPOGRAPHIC MAP SYMBOLS

VARIATIONS WILL BE FOUND ON OLDER MAPS

imary highway, hard surface
condary highway, hard surface
ght-duty road, hard or improved surface
improved road
ad under construction, alinement known
oposed road
al highway, dividing strip 25 feet or less
al highway, dividing strip exceeding 25 feet
ail

ilroad: single track and multiple track
ilroads in juxtaposition
arrow gage: single track and multiple track
ilroad in street and carline
idge: road and railroad
awbridge: road and railroad
otbridge
nnel: road and railroad
verpass and underpass
nall masonry or concrete dam
am with lock
am with road
anal with lock

uildings (dwelling, place of employment, etc.)
chool, church, and cemetery — Cem
uildings (barn, warehouse, etc.)
ower transmission line with located metal tower
elephone line, pipeline, etc. (labeled as to type)
ells other than water (labeled as to type) — Oil, Gas
anks: oil, water, etc. (labeled only if water) — Water
ocated or landmark object; windmill
pen pit, mine, or quarry; prospect
haft and tunnel entrance

orizontal and vertical control station:
- Tablet, spirit level elevation — BM △ 5653
- Other recoverable mark, spirit level elevation — △ 5455

orizontal control station: tablet, vertical angle elevation — VABM △ 95/9
- Any recoverable mark, vertical angle or checked elevation — △3775

ertical control station: tablet, spirit level elevation — BM X 957
- Other recoverable mark, spirit level elevation — X 954

pot elevation — x 7369 x 7369
ater elevation — 670 670

Boundaries: National
- State
- County, parish, municipio
- Civil township, precinct, town, barrio
- Incorporated city, village, town, hamlet
- Reservation, National or State
- Small park, cemetery, airport, etc.
- Land grant

Township or range line, United States land survey
Township or range line, approximate location
Section line, United States land survey
Section line, approximate location
Township line, not United States land survey
Section line, not United States land survey
Found corner: section and closing
Boundary monument: land grant and other
Fence or field line

Index contour
Intermediate contour
Supplementary contour
Depression contours
Fill
Cut
Levee
Levee with road
Mine dump
Wash
Tailings
Tailings pond
Shifting sand or dunes
Intricate surface
Sand area
Gravel beach

Perennial streams
Intermittent streams
Elevated aqueduct
Aqueduct tunnel
Water well and spring
Glacier
Small rapids
Small falls
Large rapids
Large falls
Intermittent lake
Dry lake bed
Foreshore flat
Rock or coral reef
Sounding, depth curve — 10
Piling or dolphin
Exposed wreck
Sunken wreck
Rock, bare or awash; dangerous to navigation

Marsh (swamp)
Submerged marsh
Wooded marsh
Mangrove
Woods or brushwood
Orchard
Vineyard
Scrub
Land subject to controlled inundation
Urban area

1 0 1 2 3 4 5 MILES

MILE SCALE 1:62 500

UNITED STATES
DEPARTMENT OF THE INTERIOR
GEOLOGICAL SURVEY

TOPOGRAPHIC
MAP INFORMATION AND SYMBOLS
MARCH 1978

QUADRANGLE MAPS AND SERIES

Quadrangle maps cover four-sided areas bounded by parallels of latitude and meridians of longitude. Quadrangle size is given in minutes or degrees.

Map series are groups of maps that conform to established specifications for size, scale, content, and other elements.

Map scale is the relationship between distance on a map and the corresponding distance on the ground.

Map scale is expressed as a numerical ratio and shown graphically by bar scales marked in feet, miles, and kilometers.

NATIONAL TOPOGRAPHIC MAPS

Series	Scale	1 inch represents	1 centimeter represents	Standard quadrangle size (latitude-longitude)	Quadrangle area (square miles)
7½-minute	1:24,000	2,000 feet	240 meters	7½×7½ min.	49 to 70
7½×15-minute	1:25,000	about 2,083 feet	250 meters	7½×15 min.	98 to 140
Puerto Rico 7½-minute	1:20,000	about 1,667 feet	200 meters	7½×7½ min.	71
15-minute	1:62,500	nearly 1 mile	625 meters	15×15 min.	197 to 282
Alaska 1:63,360	1:63,360	1 mile	nearly 634 meters	15×20 to 36 min.	207 to 281
Intermediate	1:100,000	nearly 1.6 miles	1 kilometer	30×60 min.	1568 to 2240
U. S. 1:250,000	1:250,000	nearly 4 miles	2.5 kilometers	1°×2° or 3°	4,580 to 8,669
U. S. 1:1,000,000	1:1,000,000	nearly 16 miles	10 kilometers	4°×6°	73,734 to 102,759
Antarctica 1:250,000	1:250,000	nearly 4 miles	2.5 kilometers	1°×3° to 15°	4,089 to 8,336
Antarctica 1:500,000	1:500,000	nearly 8 miles	5 kilometers	2°×7½°	28,174 to 30,462

CONTOUR LINES SHOW LAND SHAPES AND ELEVATION

The shape of the land, portrayed by contours, is the distinctive characteristic of topographic maps.

Contours are imaginary lines following the ground surface at a constant elevation above or below sea level.

Contour interval is the elevation difference represented by adjacent contour lines on maps.

Contour intervals depend on ground slope and map scale. Small contour intervals are used for flat areas; larger intervals are used for mountainous terrain.

Supplementary dotted contours, at less than the regular interval, are used in selected flat areas.

Index contours are heavier than others and most have elevation figures.

Relief shading, an overprint giving a three-dimensional impression, is used on selected maps.

Orthophotomaps, which depict terrain and other map features by color-enhanced photographic images, are available for selected areas.

COLORS DISTINGUISH KINDS OF MAP FEATURES

Black is used for manmade or cultural features, such as roads, buildings, names, and boundaries.

Blue is used for water or hydrographic features, such as lakes, rivers, canals, glaciers, and swamps.

Brown is used for relief or hypsographic features—land shapes portrayed by contour lines.

Green is used for woodland cover, with patterns to show scrub, vineyards, or orchards.

Red emphasizes important roads and is used to show public land subdivision lines, land grants, and fence and field lines.

Red tint indicates urban areas, in which only landmark buildings are shown.

Purple is used to show office revision from aerial photographs. The changes are not field checked.

INDEXES SHOW PUBLISHED TOPOGRAPHIC MAPS

Indexes for each State, Puerto Rico and the Virgin Islands of the United States, Guam, American Samoa, and Antarctica show available published maps. Index maps show quadrangle location, name, and survey date. Listed also are special maps and sheets, with prices, map dealers, Federal distribution centers, and map reference libraries, and instructions for ordering maps. Indexes and a booklet describing topographic maps are available free on request.

HOW MAPS CAN BE OBTAINED

Mail orders for maps of areas east of the Mississippi River, including Minnesota, Puerto Rico, the Virgin Islands of the United States, and Antarctica should be addressed to the Branch of Distribution, U. S. Geological Survey, 1200 South Eads Street, Arlington, Virginia 22202. Maps of areas west of the Mississippi River, including Alaska, Hawaii, Louisiana, American Samoa, and Guam should be ordered from the Branch of Distribution, U. S. Geological Survey, Box 25286, Federal Center, Denver, Colorado 80225. A single order combining both eastern and western maps may be placed with either office. Residents of Alaska may order Alaska maps or an index for Alaska from the Distribution Section, U. S. Geological Survey, Federal Building-Box 12, 101 Twelfth Avenue, Fairbanks, Alaska 99701. Order by map name, State, and series. On an order amounting to $300 or more at the list price, a 30-percent discount is allowed. No other discount is applicable. Prepayment is required and must accompany each order. Payment may be made by money order or check payable to the U. S. Geological Survey. Your ZIP code is required.

Sales counters are maintained in the following U. S. Geological Survey offices, where maps of the area may be purchased in person: 1200 South Eads Street, Arlington, Va.; Room 1028, General Services Administration Building, 19th & F Streets NW, Washington, D. C.; 1400 Independence Road, Rolla, Mo.; 345 Middlefield Road, Menlo Park, Calif.; Room 7638, Federal Building, 300 North Los Angeles Street, Los Angeles, Calif.; Room 504, Custom House, 555 Battery Street, San Francisco, Calif.; Building 41, Federal Center, Denver, Colo.; Room 1012, Federal Building, 1961 Stout Street, Denver Colo.; Room 1C45, Federal Building, 1100 Commerce Street, Dallas, Texas; Room 8105, Federal Building, 125 South State Street, Salt Lake City, Utah; Room 1C402, National Center, 12201 Sunrise Valley Drive, Reston, Va.; Room 678, U. S. Court House, West 920 Riverside Avenue, Spokane, Wash.; Room 108, Skyline Building, 508 Second Avenue, Anchorage, Alaska; and Federal Building, 101 Twelfth Avenue, Fairbanks, Alaska.

Commercial dealers sell U. S. Geological Survey maps at their own prices. Names and addresses of dealers are listed in each State index.

INTERIOR—GEOLOGICAL SURVEY, RESTON, VIRGINIA—1978

MILE SCALE 1:24 000

2 MILES 1 0 .5 1

FOOT SCALE 1:24 000

1000 0 5000 10 000 15 000 FEET

FOOT SCALE 1:62 500

5000 0 5000 10 000 15 000 20 000 25 000 FEET

the "X" on figure 8-2 has an elevation of greater than 300 feet, but less than 320 feet. It is incorrect to estimate the elevation as being about 310 feet because the X appears to lie halfway between the two adjacent contour lines.

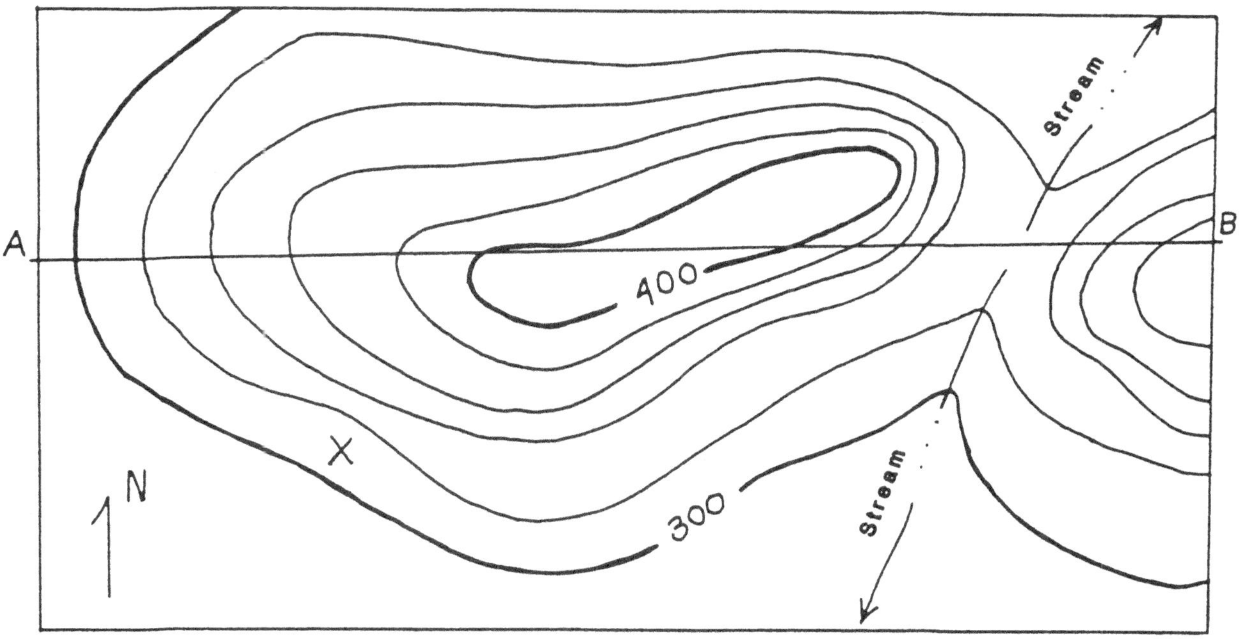

Figure 8-2. A topographic map showing an east-trending hill cut by a north-trending valley. Contour interval equals 20 feet.

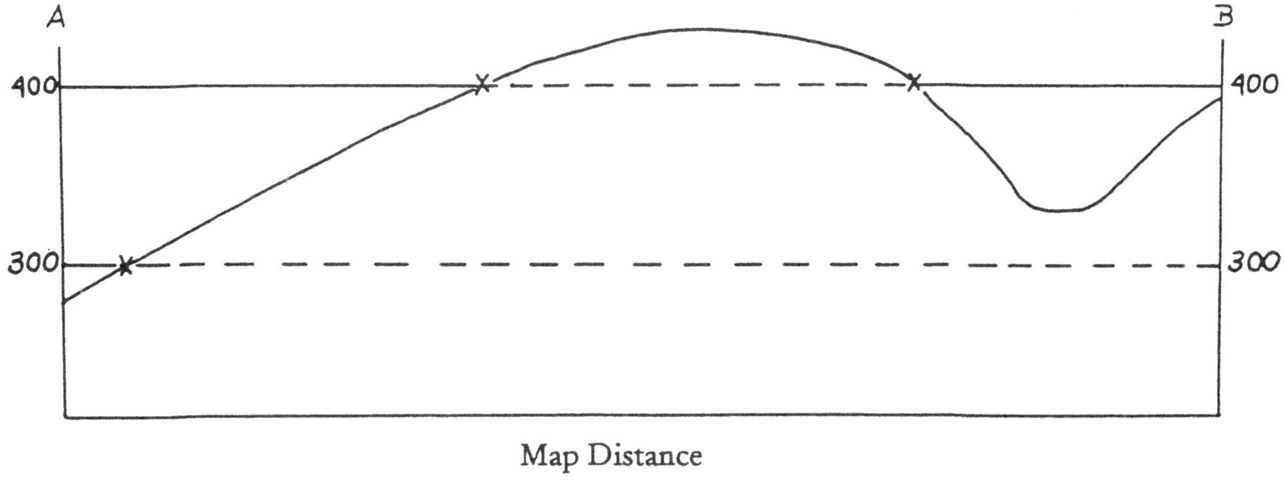

Figure 8-3. Topographic profile constructed along line A-B of figure 8-2 showing two hills separated by a valley in which two streams originate.

TOPOGRAPHIC MAP READING

A series of general rules are described and illustrated below that may aid the student in reading and using topographic maps.

1. On a steep slope, contour lines are close together.
2. On a gentle slope, contour lines are farther apart.
3. On level ground, there are no contour lines.
4. Contour lines cannot cross, except in a few rare cases, such as overhanging cliffs.
5. Contour lines must bend upstream where they cross a stream valley.

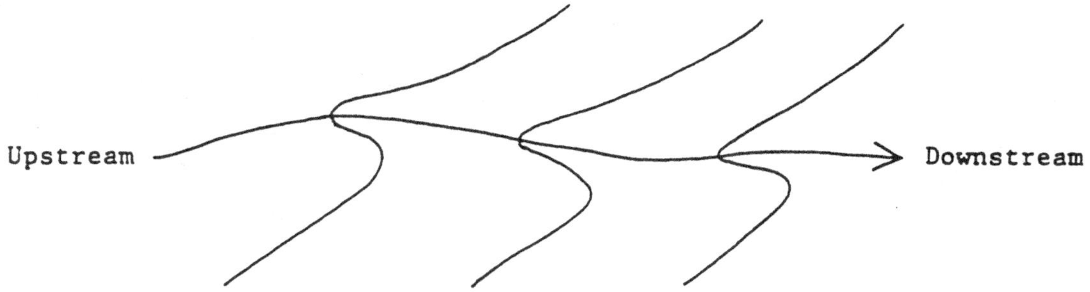

6. All points enclosed by normal contour lines are at a higher elevation than that line.
7. Closed basins are indicated by special contours known as <u>depression contour lines</u>. They are hatchered with the hatcher marks pointing downhill. All points enclosed by depression contour lines are lower in elevation than that line.

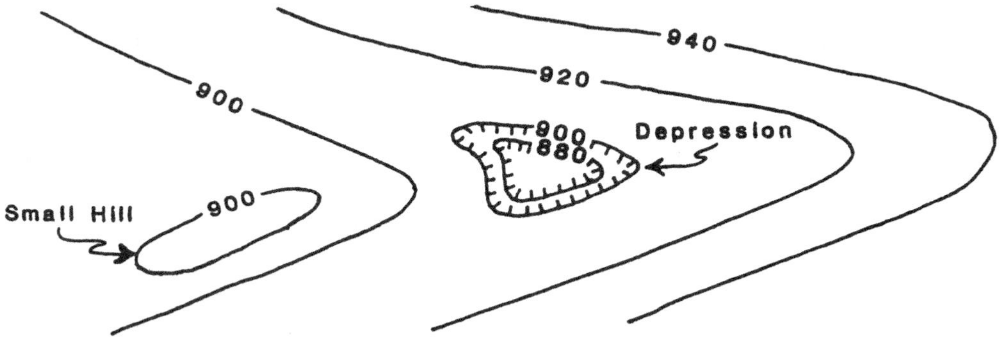

8. The outermost depression contour line has the same value as the next lower normal contour line. See above diagram.
9. Elevations of points between contour lines must be reported as ranging between the elevation values of the contour lines directly above and below.

TOPOGRAPHIC PROFILE

A topographic profile is a silhouette of a land form as seen against the sky. Profiles of land forms may be constructed from topographic maps. To do this, a specific line of profile must be selected on a topographic map. The line may cross features of interest, such as a stream valley or a hill. After selection of the line of profile, the following steps should be followed:

1. Fold a piece of paper so as to make a straight edge.
2. Place the straight side of the paper along the line of profile on the topographic map.
3. Mark on the paper (not on the map) the exact place where each contour line meets the blank straight edge on your folded paper.
4. Number each mark with the elevation of its contour line. Indicate on the paper important geographic points crossed, such as the position of streams.
5. Transfer the elevation values to a profile graph so that each point of elevation is on its proper line.
6. Connect the transferred points on the profile graph with a smooth line to complete the profile. Figure 8-4 illustrates this process.

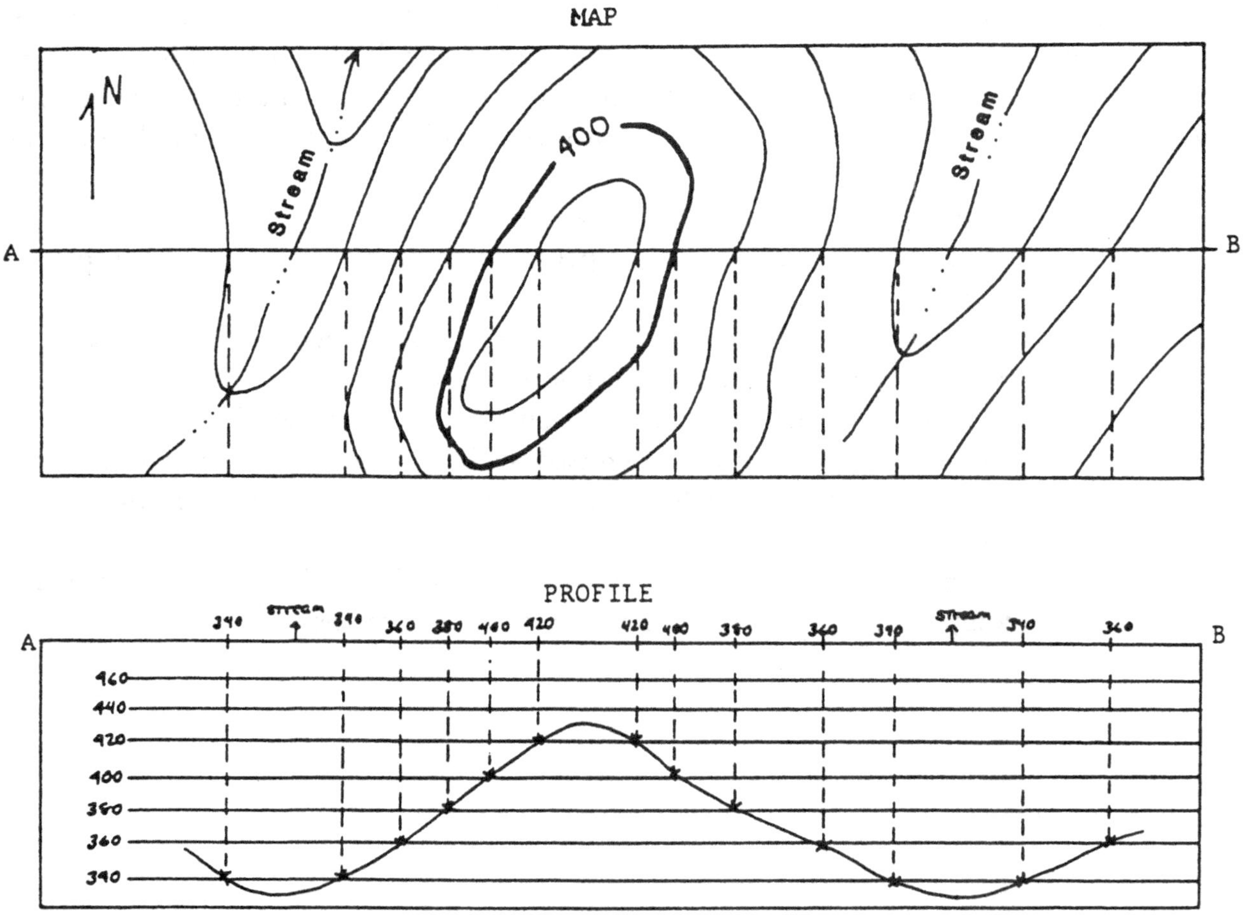

Figure 8-4. Construction of a topographic profile. A-B (upper diagram) is the line of profile on the map. The edge of the paper (lower diagram) should be placed directly along the line of profile, and the contour line values marked.

EXERCISE A

The Fayetteville, Arkansas Quadrangle covers all of Fayetteville and some surrounding areas. A study of the map will acquaint the student with many useful aspects of this map that are common to all topographic maps. Answer the following questions with respect to the Fayetteville Quadrangle:

1. Determine:
   a. Contour interval ________________
   b. Fractional scale ________________
   c. Elevation difference between index contours ________________

2. Locate "Old Main" and determine the general direction from Old Main to the following features:
   a. Farmington ________________
   b. Mount Comfort ________________
   c. Drake Airport ________________

3. Use of contour lines to determine flow direction of streams and rivers. In what general direction do the following streams flow?
   a. Cato Spring's Branch ________________
   b. Hamstring Creek ________________
   c. Town Branch ________________

4. Use of map scales to determine distances between objects on a topographic map. Determine the distances between:
   a. Old Main and Jefferson School ________________
   b. Leverett School and Mt. Comfort ________________
   c. Round Top Mtn. And Drake Field Tower ________________

5. Use of contour lines to determine elevations. Determine the elevation of the following objects:
   a. Mt. Comfort Church (exact) ________________
   b. Top of Kessler Mtn. (exact) ________________
   c. U of A Football Field (approximate) ________________

6. Locate the following features by using the Public Land Surveying System (Township, Range, Section and Quarter Section):
   a. Mt. Comfort Church ________________

b. Fayetteville Country Club Buildings ______________________________

7. Answer the following questions by referring to the Fayetteville Quadrangle:
a. Which side of Kessler Mtn, is steeper, east or west? ____________________
b. How deep would a well be if drilled on the 1800' contour line on Kessler Mtn. with its bottom at the same elevation as Mt. Zion Church (section 19)?______

EXERCISE B

This part of the exercise is designed to help the student see the relationship between contour lines and topography. Elevation points are given on the map below (figure 8-5) along with the position of a stream and its tributaries. A contour map is to be constructed directly on this map. Contour lines should be drawn in pencil, with each index contour being marked by a heavy line complete with elevation numbers. Use the contour interval given by your laboratory instructor, and hand the completed map to your instructor at the start of next week's lab.

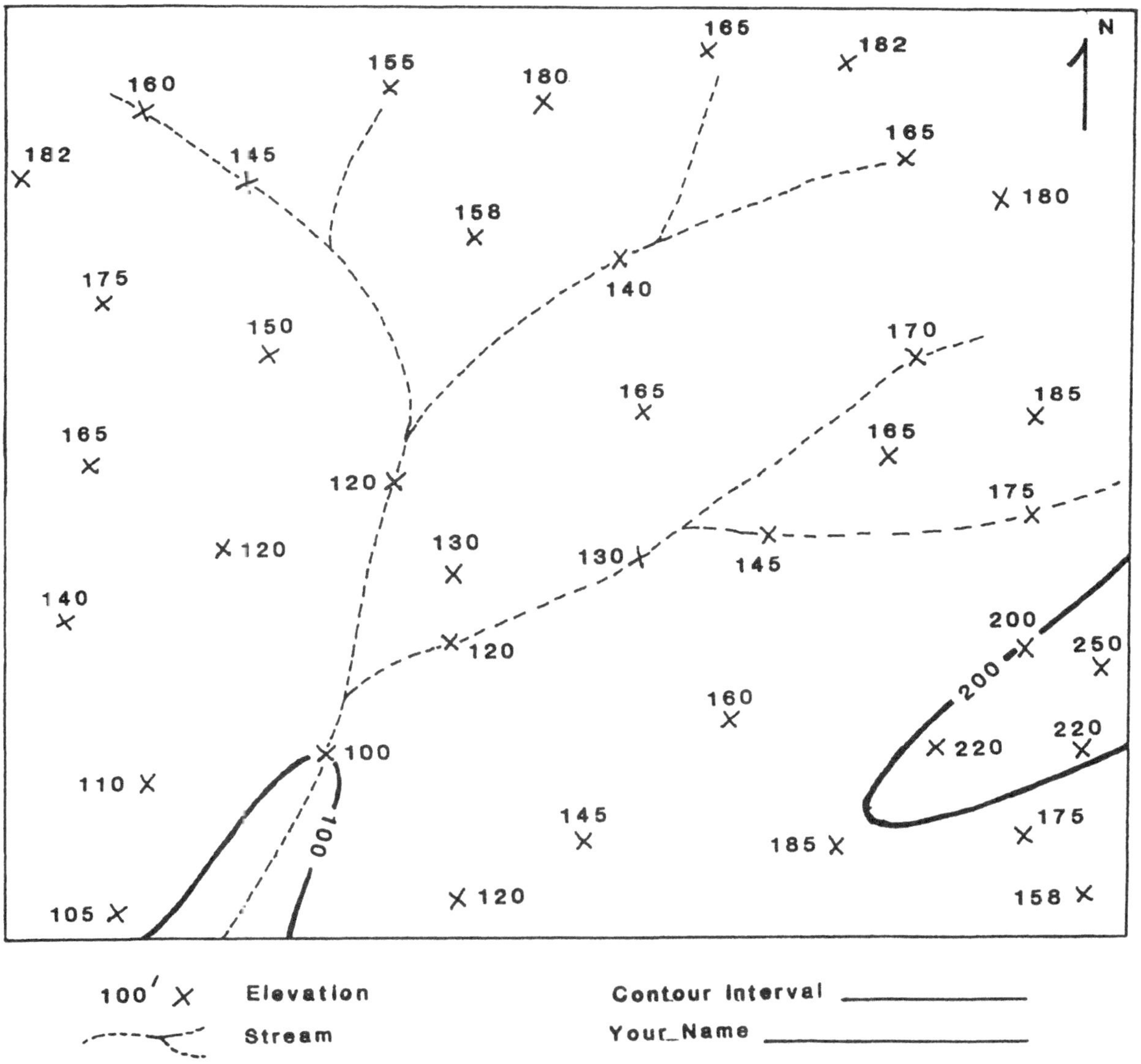

Figure 8-5. Rectangular area showing the distribution of streams and points of known elevation.

EXERCISE C

This part of the exercise is designed to help the student see the relationship between normal contour lines, depression contour lines, and topography. Contour lines are provided on figure 8-6. Label each contour in the space provided. A contour interval of 20 feet has been used. Darken all index contours and use standard intervals such as 700, 720, 740, 760, for the contour lines. A single elevation is shown at 735 feet. Streams are indicated by dark lines. Complete this exercise at home and hand it to your instructor at the start of next week's lab.

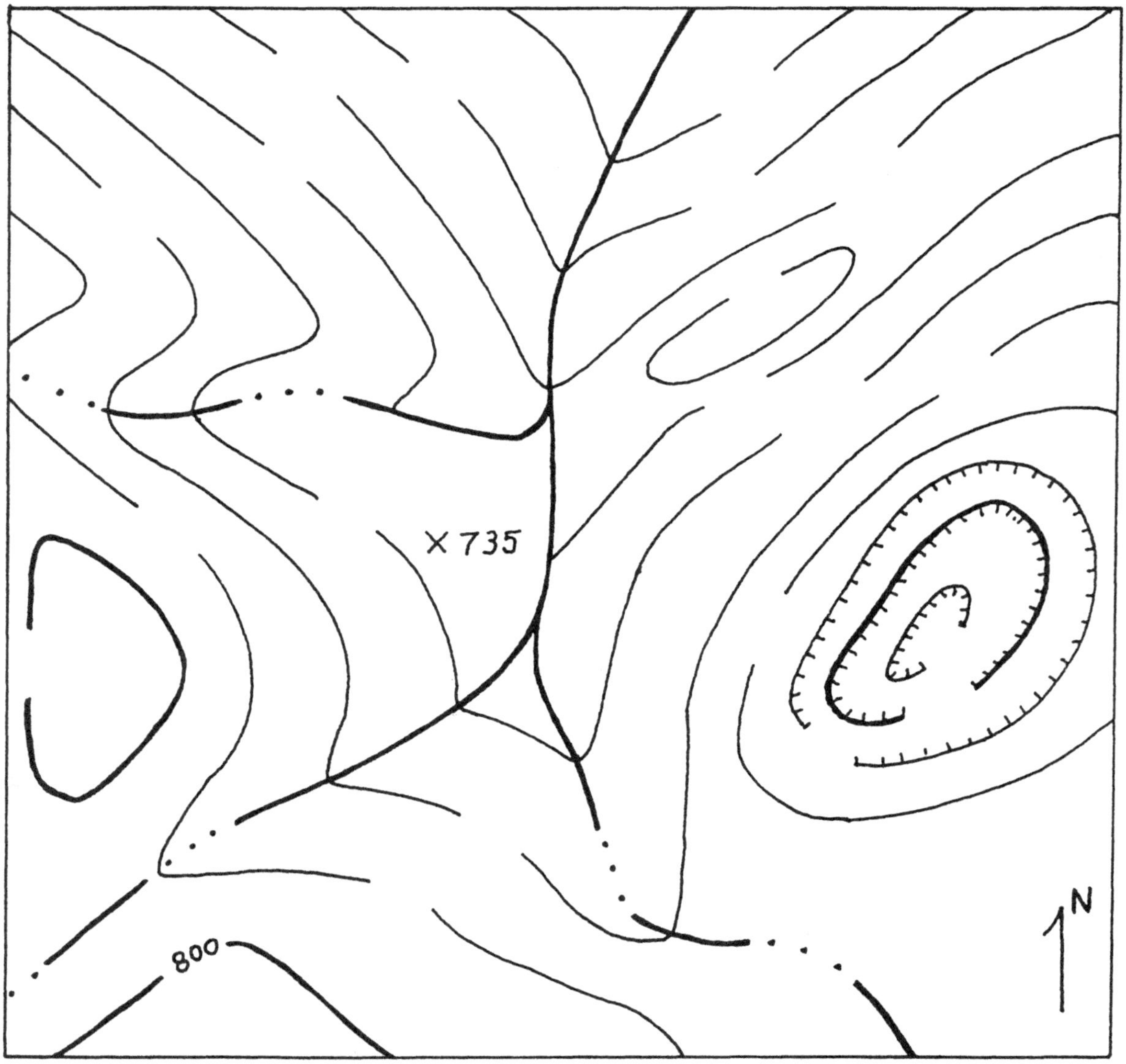

Figure 8-6. Topographic map showing normal contours, depression contours, and the location of streams (heavy black lines).

EXERCISE D

The object of this exercise is to construct a topographic profile from the Fayetteville Country Club Northwest to the water tower near the summit of Mt. Sequoyah. Use a separate sheet of paper to locate and determine elevation points. Transfer these elevation points to the profile graph given below and draw the profile. Complete this problem in lab today.

Figure 3-7. Graph paper for constructing a topographic profile from Fayetteville Country Club to the water tower on Mt. Sequoyah.

CHAPTER 9

GEOLOGIC INTERPRETATION OF REMOTE SENSOR IMAGERY AND TOPOGRAPHIC MAPS

by

ROY VANARSDALE

Topographic maps and remote sensor imagery allow geologists to rapidly examine large areas of the Earth's surface. Landscape features and the relationships that exist between them, obscure to an individual on the ground, may be detected and defined. Subtle geologic structures are often clearly expressed by maps and photographs. Because of this, the use of maps and remote sensor images has become an increasingly important geologic tool.

In this exercise, we are going to examine types of landscapes in terms of the specific features that compose them and the geologic processes that were important in their formation.

STREAM EROSION AND DEPOSITION

Streams are active erosional agents in both humid and arid regions and are responsible for most of the landscape features of the Earth's surface. Specific landscape features and concepts important in their definition are considered below.

V-shaped valleys In upland areas, the erosive energy of streams is directed downward against the stream bed. Continued erosional activity of this sort will produce deep valleys. The walls of these valleys taper downward toward the stream channel producing a V-shaped valley.

Graded stream When a stream exhibits a balance between its ability to transport and deposit sediment it is graded. Waterfalls, rapids and sharp V-shaped valleys are absent and a smooth longitudinal profile develops.

Flood plain valleys When a stream becomes graded some of the erosive energy of streams is directed laterally against the valley walls. As the walls retreat, the valley becomes extremely broad, and the channel occupies only a small part of the entire valley.

Flood plain Broad flat areas that develop adjacent to stream channels are called flood plains. These plains are subjected to periodic flooding during times of heavy rain when the channel cannot contain the flow of the stream. Sediment carried from the channels onto the flood plain is deposited as alluvial material on a broad flat surface.

Meandering stream pattern Stream channels on flood plains are often characterized by broad sinuous curves called meanders. Meanders are produced as the stream channel migrates toward the outside of a meander bend. The current in the channel erodes the outside of this bank, called the cut bank, because the current velocity is higher in this area. The inside bank of a meander curve is the site of sediment deposition because current velocities are low. This bank is the slip-off slope. Since the current continually erodes the cut bank and forms sedimentary deposits on the slip-off slope, the stream channel migrates laterally, accentuating the meander pattern.

As the channel migrates, continued deposition on the slip-off slope produces a characteristic deposit within the bend of the meander. This deposit is called a point bar deposit (figure 9-1) and contributes to the construction of the flood plain (figure 9-2).

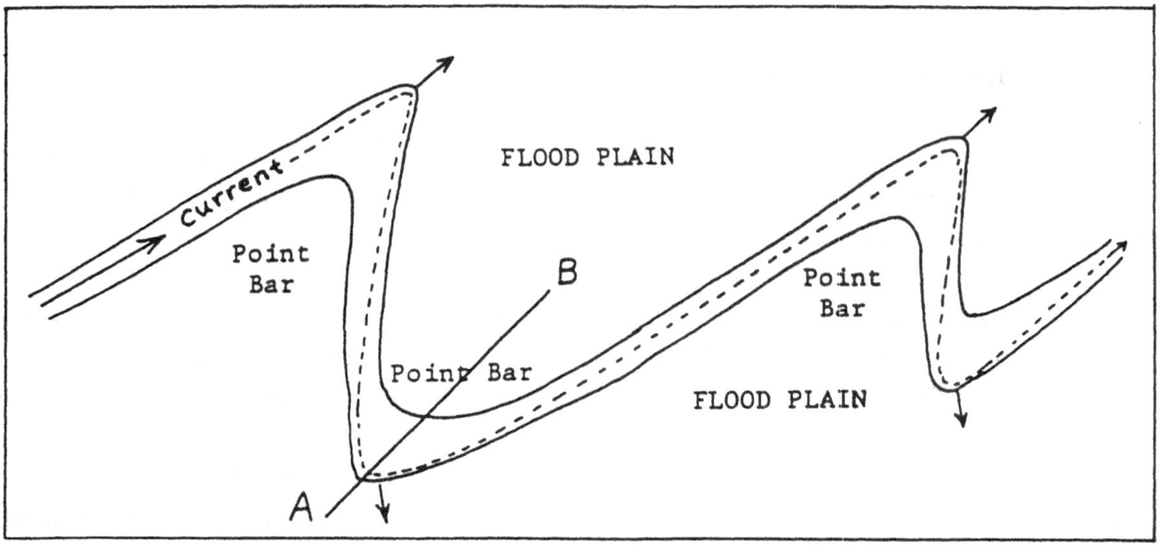

Figure 9-1. Map view of a meandering stream showing the location of point bar deposits and associated flood plain. Arrows show migration direction of individual meander loops.

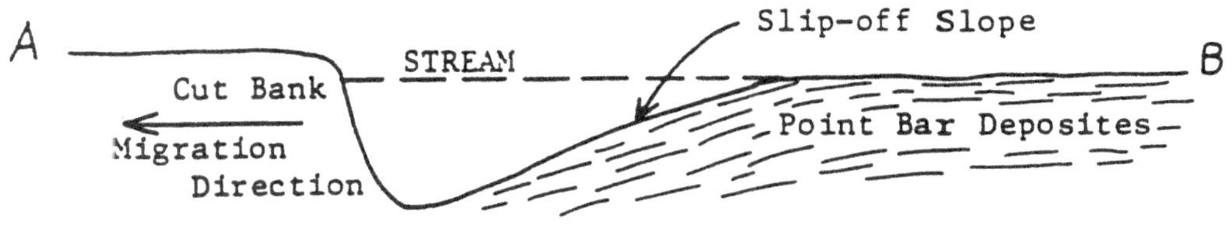

Figure 9-2. Cross sectional sketch drawn along line A-B of figure 9-1 showing the relationship between a cut bank, slip-off slope and point bar deposits.

Frequently, highly accentuated meanders are cutoff by flood waters that erode a new channel across the older point bar deposits. The ends of the old channel are filled with sediment and the former meander bend is isolated from the active channel. This process produces oxbow lakes, a characteristic feature of many flood plains (figure 9-3). As oxbow lakes are slowly infilled with sediment, they lose their lake characteristics and become shallow topographic lows known as meander scars.

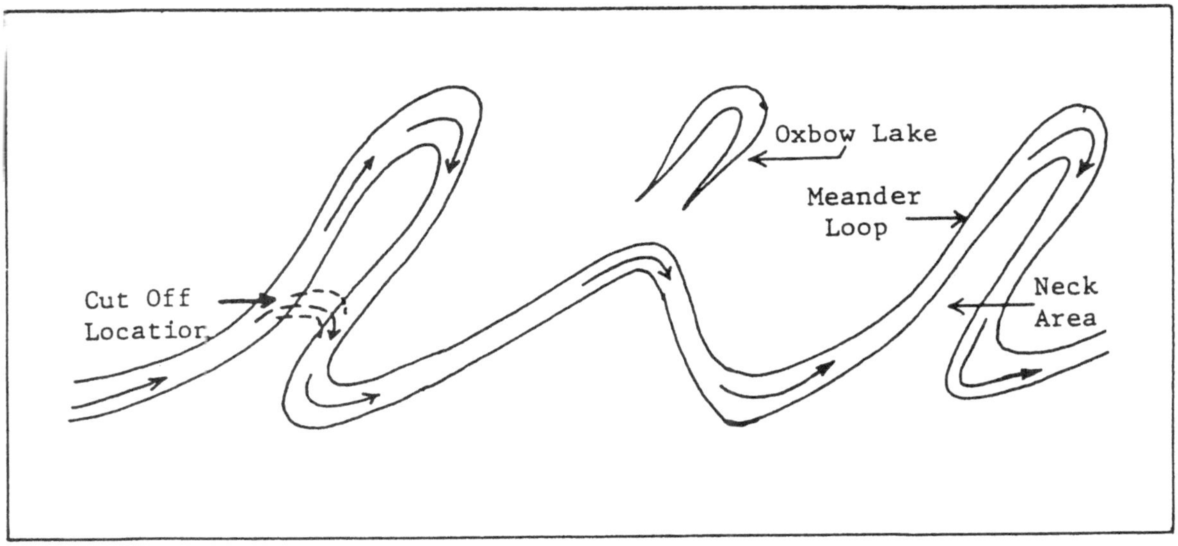

Figure 9-3 Map view of a meandering stream showing a meander loop, location of a possible cutoff in a neck area, and an oxbow lake.

Flood waters, when they exceed the carrying capacity of the channel, spread across the flood plain and deposit sediment. Currents rapidly diminish in velocity as the channel banks are overtopped and much of the sediment carried by the water is deposited immediately next to the channel. This produces low ridges adjacent to the channel called natural levees (figure 9-4).

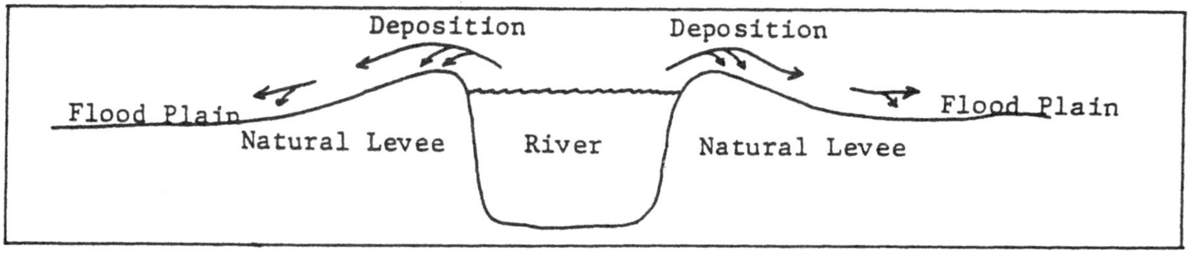

Figure 9-4. Cross section diagram of an alluvial stream channel showing a flood plain and two natural levees.

Braided stream pattern Streams that are heavily laden with sediment and are unable to carry their load develop numerous channels that divide and rejoin around small islands. Such stream patterns often occur in front of a melting glacier where vast quantities of sediment are available. Under these conditions the stream is simply not able to transport all the material. Sediment is deposited in the channel forming the islands, and the stream divides and passes around them. Braided streams are also formed in arid and semi-arid areas when inadequate rainfall does not produce flow sufficient to move available sediment (figure 9-5).

Dendritic drainage pattern Streams that flow over rock material with uniform chemical and physical properties develop tributaries that have a pattern resembling that of the branches of a tree. This pattern is referred to as dendritic and forms on uniformly erodible sedimentary rock and on unfractured igneous rock associated with batholiths (figure 9-5).

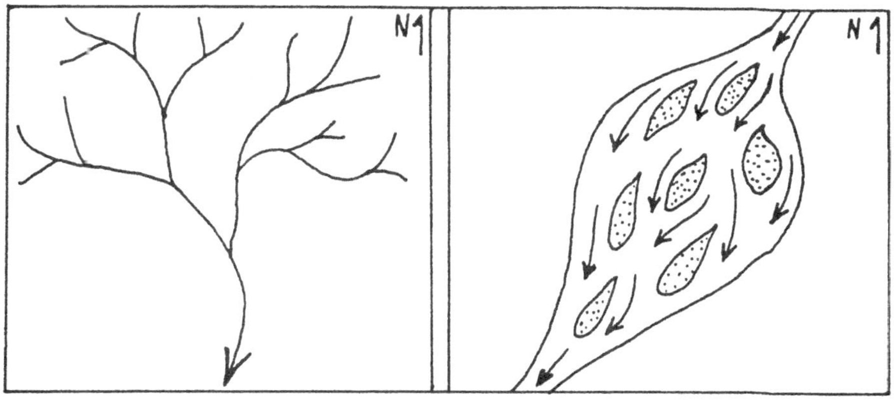

Figure 9-5. Map view of a dendritic drainage pattern (left side) and a braided stream pattern (right side).

Gradient The gradient of a stream is simply the steepness of the slope down which a stream flows. It is measured in feet of fall per mile of travel. To determine gradient along a particular stretch of a stream, measure the length of the stretch in miles, then determine the elevation difference between the upper end of the stream and the lower end of the stream in feet using the topographic contours. Divide the length of the stream in miles into the total number of feet of fall to determine the gradient in feet per mile. The gradient of a stream is often an indicator of the ability of the stream to erode and transport sediment.

Alluvial fan An alluvial fan is an apron of sediment deposited at the foot of mountains in arid lands. Streams flowing from the mountains flow out onto the flat desert floor. The stream's gradient and thus its velocity diminishes rapidly. It becomes unable to transport all the available sediment and deposition occurs. Braided stream patterns often develop on alluvial fans in response to the steams inability to transport all available sediment.

Landforms The landscapes around us represent a balance between the geology (rock type and structure) at the Earth's surface and the erosional forces (running water, ice, and wind) which are working to erode the landscape. The angle of hill slopes and the gradient of rivers are largely controlled by the resistance (strength) of the underlying rock. For example, gentle hill slopes form on weak shales and steep hill slopes form on strong sandstones. Streams form wide flood plains where they flow across weak rock and have narrow flood plains where they flow across strong rock. Due to the relationship between landscape and the underlying geology, we are able to interpret the geology which is responsible for the landforms which we see around us. Figure 9-6 illustrates how the resistance and orientation of bedrock controls certain topographic features. In figure 9-6, layered sedimentary rock in the Earth's crust has been deformed by bending (folding). Streams tend to rapidly remove weak (nonresistant) rock, like shale, to produce valleys. Resistant rock, like sandstone, is eroded more slowly and tends to stand high as plateaus or ridges terminated by steep cliffs often referred to as escarpments. Basic landforms related to the resistance and/or orientation of bedrock are hogback ridges, cuestas, buttes and mesas. These topographic features are defined as follows:

1. Hogback Ridge Symmetrical ridge produced by steeply inclined resistant rock.

2. Cuesta Asymmetrical ridge produced by gently inclined resistant layers. The layers are tilted in the direction of the gentle slope.

3. Butte and Mesa Irregular flat topped hill bounded by alternating cliffs and more gentle slopes. The gentle slopes reflect less resistant rock, like shale, and the steep slopes are produced by resistant layers. The rock layers are horizontal. A butte is considered to be a small mesa by most Geologists.

4. Plateau A broad, elevated, flat-topped area normally developed on nearly horizontal strata, and terminated by a pronounced escarpment.

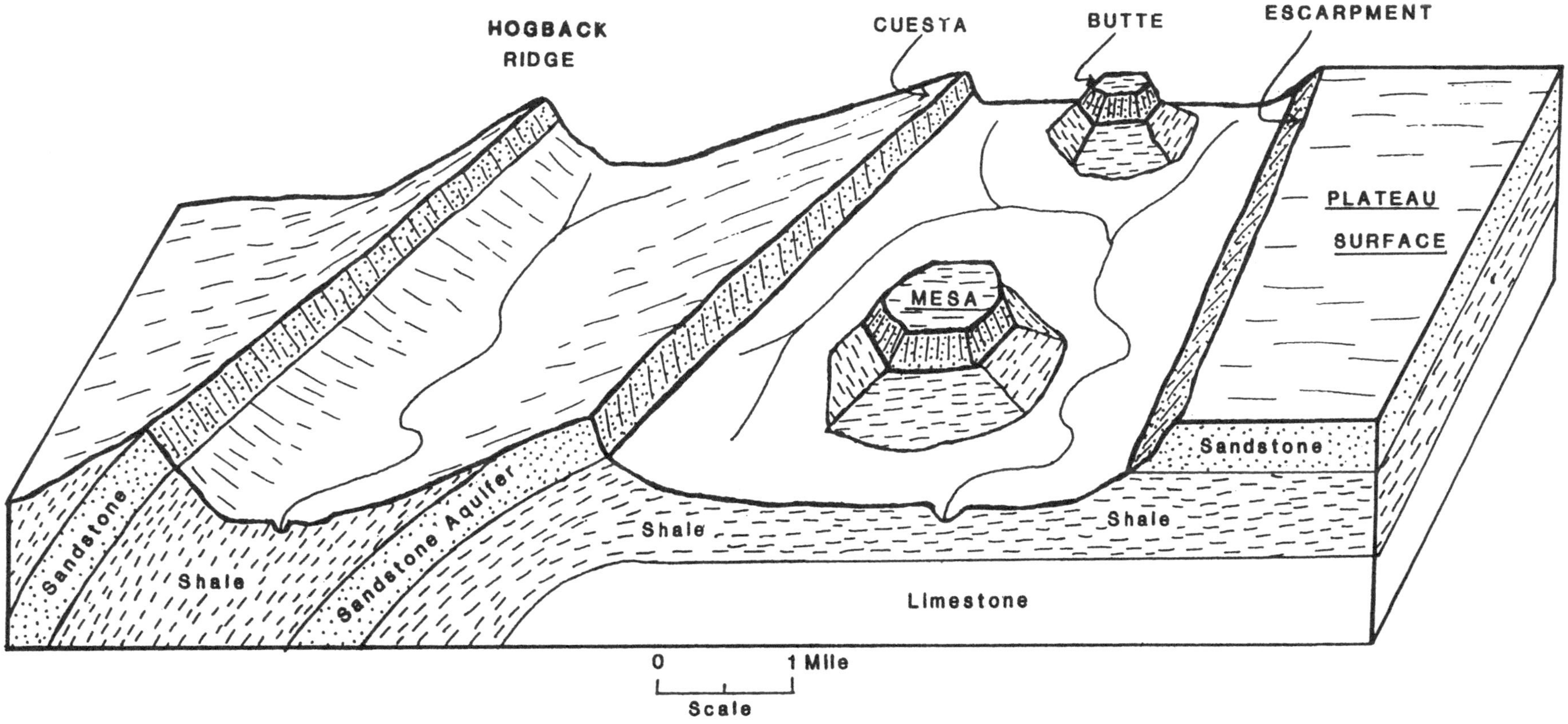

Figure 9-6. The relationship between hogback ridges, cuestas, mesas, buttes and a plateau surfaces.

EXERCISE A

The Fayetteville, Arkansas Quadrangle covers an area where flat lying sedimentary rocks are 300 to 350 million years old. These rocks include limestone, shale, siltstone, and sandstone and have been assigned to the Boone, Hindsville, Batesville, Fayetteville, Pitkin, Hale, Boyd, and Atoka Formations. Sandstone of the Atoka Formation and limestone of the Pitkin Formation (where present) are particularly good bluff formers. Answer the following questions with respect to the Fayetteville Quadrangle.

1. Describe the topography that is characteristic of the southeastern part of the map.

   ____

   ____

2. Describe the topography of the northwestern one-third of the map.

   ____

   ____

3. Determine the gradient of Cato Springs Branch from its headwaters to its intersection with the Burlington Northern Railroad. (Add one-half inch to length of river to account for turns in the river).

   ____

4. Determine the gradient of the White River from just south of Puddin Hill to where it leaves the map farther south. (Add 1 inch to length or river to account for turns in the river). ______________________________

   ______________________________.

5. Construct a topographic profile from the 1400 foot contour on Puddin Hill southwest across the White River Valley to the 1400 foot contour on the hill east of Lake Wilson. Use the profile graph provided below. Contrast this profile with one constructed from the cemetery on the western side of McCullom Mountain through the C in Cato to the 1400 foot contour on the hill west of Cato Springs Branch.

WHITE RIVER VALLEY

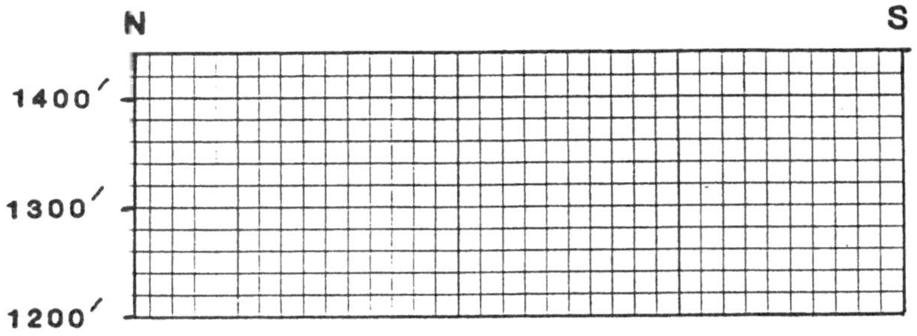

CATO SPRINGS BRANCH

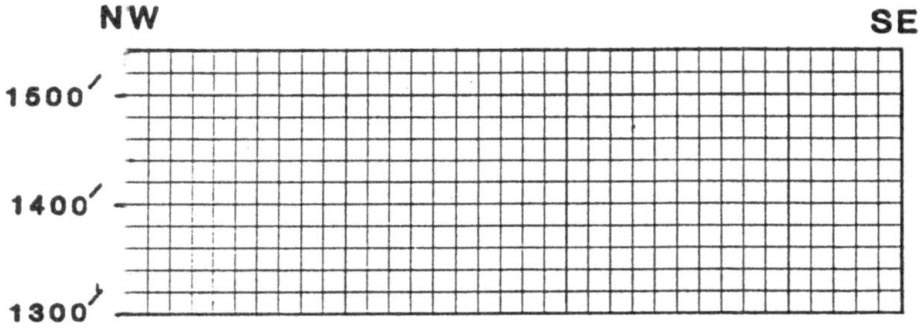

6. Why is the valley bottom of the White River flat?

____

____.

7. Would you expect to find bedrock and waterfalls and rapids along the White River channel? ____.

8. Would you say that the White River is graded ____________________, and if so how is the river expending some of its erosive energy?

____

____.

9. What name should be applied to the stream pattern shown by the White River?

____.

10. How would you describe the cross sectional shape of Cato Springs Branch?

____.

11. Would you expect to find bedrock and waterfalls and rapids along Cato Springs Branch? ______________________________.

12. Would you say that Cato Springs Branch is graded ____________, and if not, how is the stream expending most of its erosive energy? ______________________________ ______________________________.

13. Determine the average elevation of flat-topped Puddin Hill, South Mountain, and Washington Mountain. ________________. Do the same thing with the Miller and Kessler Mountains.________________.

EXERCISE B

The Prairie Grove, Arkansas Quadrangle is located immediately southwest of the Fayetteville Quadrangle. Topography in this quadrangle is also controlled by flat lying sedimentary rock. Anser the following questions with respect to the Prairie Grove Quadrangle.

1. Examine Stevenson Mountain in sec. 27, T. 15 N., R. 31W.. Determine the approximate elevation of the top of this flat topped mountain. ________________.

2. Compare the elevation of the top of this mountain with elevations for flat topped mountains in the Fayetteville Quadrangle. Do you note any similarities? ________.

   What does this mean with respect to horizontal persistence of the flat lying sedimentary rocks in the area? ______________________________ ______________________________ ______________________________

3. If there were little or no vegetation in this area, what name would surely be applied to these hills? ______________________________ ______________________________.

EXERCISE C

Examine the various aerial photographic pairs placed around the laboratory, and answer the questions given below.

1. Photograph #45 (South Dakota)

   a. Describe the cross sectional shape of the numerous small stream valleys in the northern one-half of the area. ______________________________

   b. What type of drainage patten is well developed in the badlands area to the north ____________________, and what does this tell you about the bedrock in the area? ______________________________

   c. Why are the fields in the southern one-half of the area different shades of gray? ______________________________
   ______________________________.

2. Photograph 40 (South Dakota)

   a. What type of stream pattern is illustrated by Bear Creek (major stream)?
   ______________________________.

   b. List any recognizable stream features associated with Bear Creek.
   ______________________________.

   c. Would you say this area is (densely, moderately, or sparsely) vegetated?
   ______________________________.

3. Photograph 34 (Mississippi)

   a. Examine Horseshoe Lake (B5-2.0), and then locate another oxbow lake and a meander scar using the grid system on the photograph.

   Oxbow Lake ____________ Meander Scar ________________

b. What type of material is the Tallahatchi River flowing on ________________, and what geological term should be applied to the smooth surface on either side of the river? ________________________.

c. Locate a meander neck where a new cutoff is likely to form in the future ______________________________ and a well developed natural levee ____.

4. <u>Photograph 25</u> (California)

a. Identify the three fan-shaped features shown in the middle frame of this photograph. ____.

b. Identify the numerous small mounds located in B4-3.4. ____.

In what general direction is the wind blowing in this area? ____________.

c. Examine the dark areas where bedrock is exposed and state whether the bedrock is igneous or sedimentary. ______________________________, and explain your answer. ____ ____.

EXERCISE D (for extra credit)

Examine the Waldron, Arkansas Quadrangle map, and answer the questions given below.

1. Construct a topographic profile from Bush School north to Haw Creek on the graph paper below. Use only index contour lines in your construction.

BUSH SCHOOL HAW CREEK

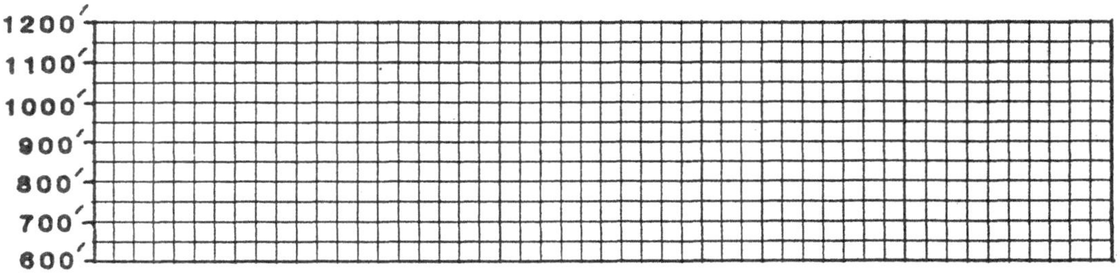

2. Draw the upper and lower boundaries of two resistant and two nonresistant rock formations on the above profile that could be responsible for this type of topography. (See figure 9-6 for an example.)

3. Examine the topographic profile you have just constructed, and compare it with figure 9-6. What name should be applied to this topographic feature?

   ____.

4. Assume that the resistant rock units in the above topographic profile are porous and permeable sandstone. Could the city of Waldron use this information to improve their water supply? ________________________________.

5. If the answer to question number 4 was yes, explain your answer below. Do the same thing if your answer to question number 4 was no. (Hint - see artesian aquifers and wells in textbook). ______________________________

   ____

   ____.

CHAPTER 10

FIELD TRIP TO THE ILLINOIS RIVER VALLEY

by

MARGARET GUCCIONE AND DOY ZACHRY, JR.

The landscape of northwestern Arkansas is dominated by three major plateaus. Their distribution is indicated on figure 10-1. The Boston Mountain plateau is the highest of the three and generally coincides with the area covered by the Boston Mountains south of Fayetteville. This plateau, originally a continuous flat surface, has been severely eroded by the White, Illinois, Buffalo, and Mulberry rivers and their tributaries. Today, the flat tops of numerous mountains are all that remain of this formerly continuous surface. The Boston Mountain plateau surface is visible to the south (left side of bus) of Highway 16 on the way to the Illinois River.

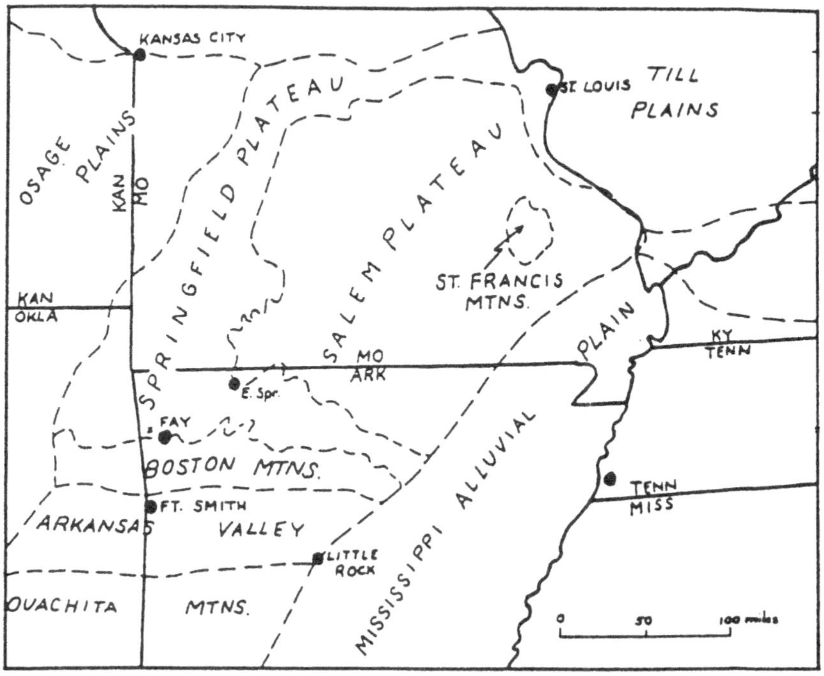

Figure 10-1. Province map showing the distribution of the Salem plateau, the Springfield plateau, and the Boston Mountains (Boston Mountain plateau). Other provinces in Arkansas and surrounding states are also shown. (After G.G. Huffman, Oklahoma geologic Survey, Bulletin 77).

The Springfield plateau is lower and lies north of the Boston Mountain plateau. This plateau has not been as severely eroded as the Boston Mountain plateau and is characteristically a gently rolling surface that continues northward from Fayetteville into Missouri. A few streams such as the White and Illinois rivers have eroded valleys into its surface.

The Salem plateau is the lowest of the three surfaces. Driving north through Eureka Springs into Missouri, one passes through the margin of the Springfield Plateau and descends to the Salem plateau. The margin of the Springfield plateau has been strongly eroded by streams in this area creating the mountainous topography at Eureka Springs.

The field trip to the Illinois River near Lake Wedington is designed to illustrate a number of landscape features associated with the Springfield plateau and with a stream valley eroded into that plateau. Highway 16 passes westward from Fayetteville along the Springfield plateau surface to the valley of the Illinois River. The landscape is gently rolling with few mountainous features in contrast to the Boston Mountains visible to the south.

The Springfield plateau surface has been exposed to weathering for a substantial time and thin soils are widely developed. These soils are often referred to as pedalfers by geologists. They contain a thin A horizon, the upper part of which is brownish because of organic material (humus), and a much thicker tan colored B horizon containing clay and some insoluble rock fragments composed of weathered parent rock from which the residual soil was derived.

The highway descends from the Springfield plateau surface into the valley of the Illinois River. The river has eroded downward through limestone and shale layers of Devonian and Mississippian age. These layers, assigned by geologist to the Chattanooga, St. Joe, and Boone Formations, are visible in the eastern valley wall. A stratigraphic section (figure 10-2) and a photograph (figure 10-3) illustrates the formations and members and their characteristic rock types.

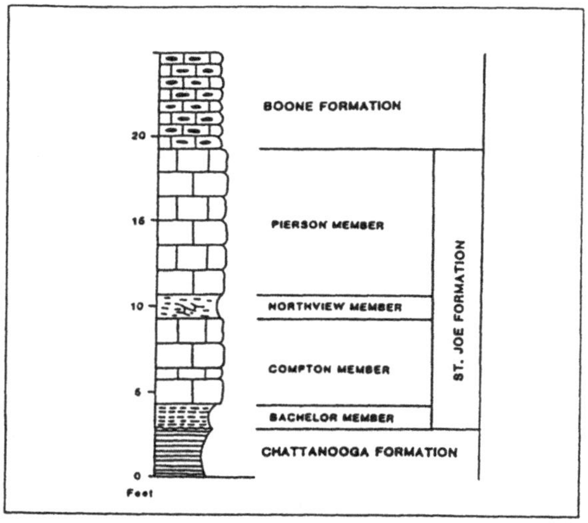

Figure 10-2. Stratigraphic units exposed on the eastern valley wall of the Illinois River. Limestone of the Boone Formation contains chert whereas the Pierson and the Compton do not. The Chattanooga Formation is of Devonian age and the St. Joe and Boone formations are of Mississippian age.

Figure 10-3. Members of the St. Joe Formation are exposed in this outcrop on the eastern valley wall of the Illinois River. The Bachelor overlies the Chattanooga Formation, but the contact is beneath the road and not visible. The Northview Member of the St. Joe Formation here is a thin unit of calcareous shale. To the northeast near Springfield, Missouri, the unit thickens to over 50 feet and is used to safely house landfills in a terrain dominated by pollution-prone limestone.

The Chattanooga Formation is composed of black shale and ranges to a thickness of about 50 feet in northwest Arkansas. Only the upper foot or so is exposed in the valley wall of the Illinois River; the rest is below the land surface (figure 10-4).

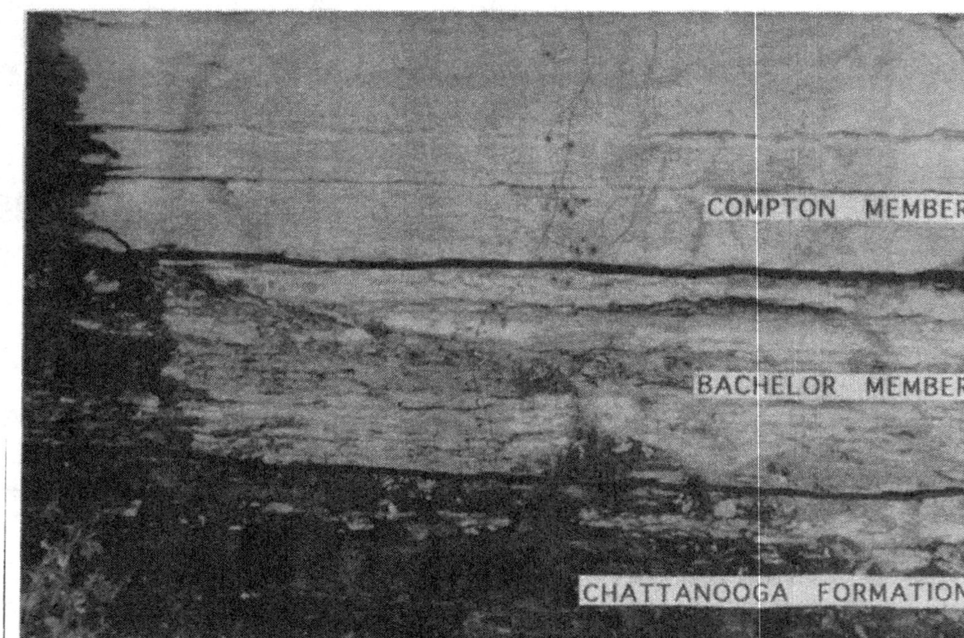

Figure 10-4. Black shale of the Chattanooga Formation overlain by brown shale of the Bachelor Member of the St. Joe Formation. Deposition of black mud stopped as the Chattanooga sea withdrew. A later Mississippian sea then advanced across the land surface and the brown shale and limestone of the St. Joe Formation was deposited. The boundary represents an ancient land surface that existed between advances of the two seas.

The Chattanooga accumulated as black mud in an extremely widespread sea that covered most of the southeastern part of the United States during Late Devonian time (about 370 million years B.P.).

The overlying St. Joe Formation also accumulated in a broad seaway that covered much of the United States at a slightly later time. Unlike the Chattanooga seas, the waters were clear and numerous invertebrate organisms thrived. Their shells accumulated on the sea floor and, in time, formed the Compton and Person Members of the St. Joe Formation (figure 10-4). Crinoids or sea lilies were particularly abundant and the calcareous stems on which they supported their bodies are frequently preserved as fossils (Figure 10-5).

Figure 10-5. Fossils of crinoids or sea lilies from the St. Joe Formation. The organisms were invertebrates that lived stationary on the sea floor and were supported above the bottom by the calcite stems seen here.

Ground water currently moves through the fractures and caves of the St. Joe and Boone Formations. Evidence of water movement appears as numerous springs that can be seen along the south side of the road leading back to Highway 16. Because limestone is soluble, the fractures continually become wider until they are able to transport large volumes of water beneath the surface. This type of ground water is often polluted because fractured limestone does not filter out contamination products as effectively as a rock like sandstone. The wide spread occurrence of limestone and dolostone in northwestern Arkansas has increased the incidence of pollution of many springs in the area in recent years.

The Illinois River Valley has many features characteristic of rivers that both erode bedrock and deposit sediment. Particularly well exposed at this stop are slip-off slopes, cut banks and point bars. The flood plain of the Illinois River is illustrated in Figure 10-6. The photograph was taken from the bridge on Highway 16 with the camera pointing east. The St. Joe Formation is visible in the eastern wall of the valley. During flooding in the 1980's. The flood plain was completely inundated and the small side road beside the St. Joe Formation was under water.

Figure 10-6. Flood plain of the Illinois River near Savoy. The photograph was taken from the bridge on Highway 16. A prominent exposure of the St. Joe Formation in the distance forms the eastern wall of the valley. During a flood in the 1980's, the entire plain was covered and open water extended to the rock exposed in the valley wall.

LABORATORY EXERCISE

1. Using the Data and Observation Form below as a guide, record observations and other information at each stop. The completed form will be examined by your instructor at the end of the trip.

2. Using the Wheeler 7.5 minutes topographic map (Figure 10-7) construct a topographic profile from A to B on graph paper, and hand it to your instructor at the start of your next laboratory class for extra credit.

DATA AND OBSERVATION FORM

NAME ______________________________

Stop 1. Illinois River Valley

A. Locate this stop by township, range and section (figure 10-7)

____.

B. How was this valley formed? ______________________________________.

C. What type channel does the Illinois River have, straight, meandering, or braided?

____.

D. At the bridge, in what part of the channel is the water deepest? __________________.

E. In what part of the channel is the velocity of the water greatest? __________________.

F. Farther downstream (to the north) finer grained sediment is exposed in the channel wall. How were these sediments formed? __________________________.

G. What name should be applied to the slightly higher, flat surface on the west side of the valley? ____.

H. What age do you think the sediment of question F might be? (Refer to the geologic time chart, Figure 5-2). ____________________________________.

Stop 2 <u>Paleozoic Rock of Northwest Arkansas</u>

A. What kinds of sedimentary rocks are exposed along the valley walls of the Illinois River and Clear Creek? ______________________.

B. Approximately how old, in millions of years, are these rocks? ____________.

C. Are these rocks older or younger than the sediments in the valley? __________.

D. What is the orientation of these rock units? ______________________.

E. An unconformity is present within the shale. There are no rocks preserved for an interval of 10 million years. Can you identify this unconformity in the field?
____.

F. Why are the tree trunks curved and tilted on slopes above the rock exposure?
____.

G. Why is subsurface water that flows only through fractures and/or caverns in carbonate rock easily polluted? ______________________
____

Stop 3 <u>Soil Developed on the Boone Limestone</u>

A. What soil horizons are present? ______________________.

B. Why is the uppermost part of the soil darker colored? ______________.

C. What type of rock is present as fragments in the soil? ______________.

D. What other type of rock was originally present but has been dissolved during weathering? ______________________.

E. Refer to figure 10-2. What formation has been weathered to produce the soil profile present at this stop? ______________________.

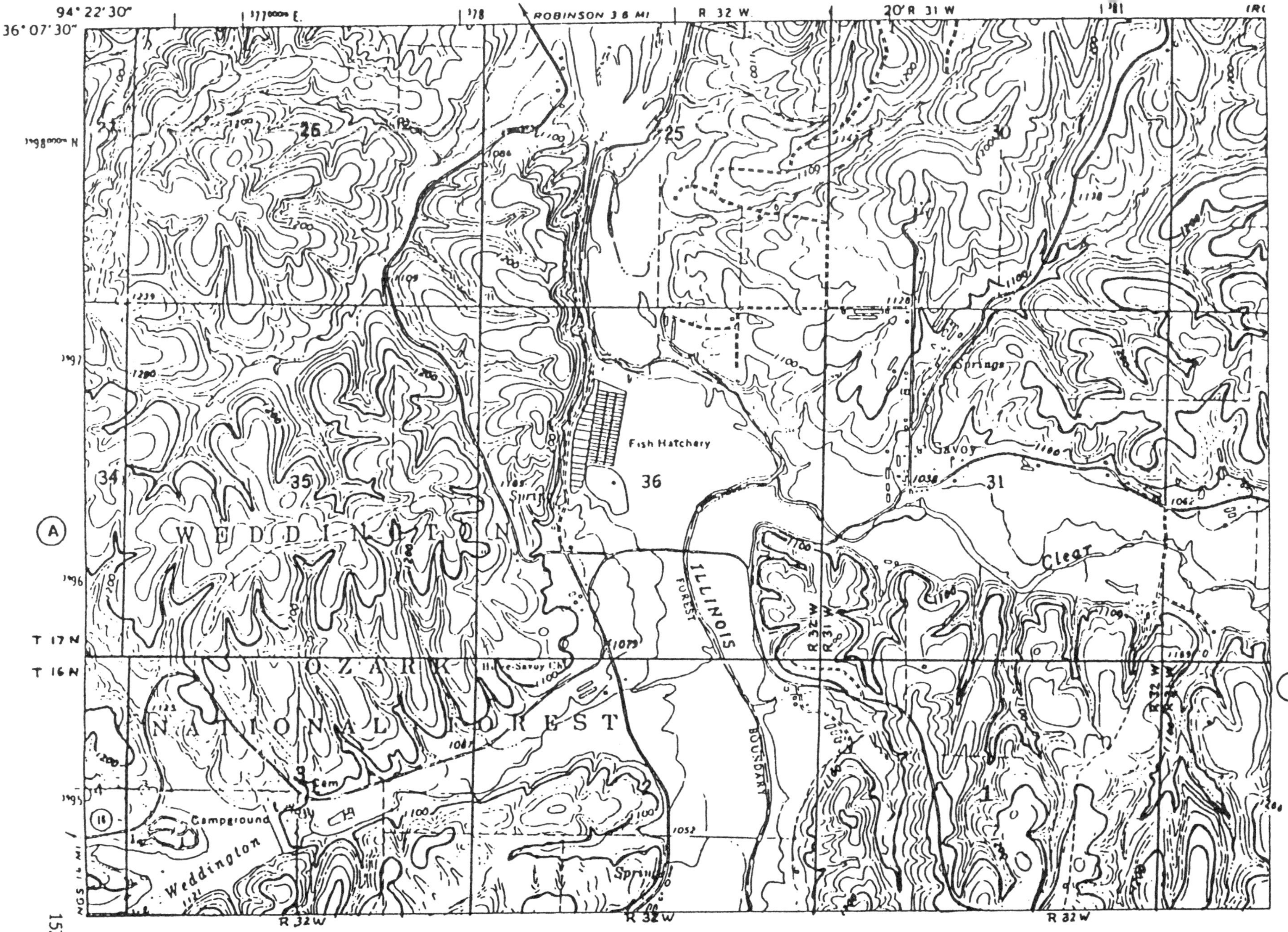

Figure 10-7. A portion of the Wheeler 7 ½ minute topographic map, Washington County Arknasas. Contour interval = 20 feet.

CHAPTER 11

GLACIERS AND GLACIATION
by
RONALD H. KONIG

INTRODUCTION

Land areas once covered by either valley ice or more extensive ice sheets are said to have been glaciated. Some of the Earth's most spectacular topography is the result of glaciation and associated mechanical weathering. Features related to continental and alpine glaciation may be seen on both topographic maps and aerial photographs. The objective of this exercise is to view topographic features in either two or three dimensions and relate them to glaciation and related activities.

EROSIONAL FEATURES OF ALPINE GLACIATION

Prior to glacial activity, a mountain range often consists of somewhat rounded peaks and a series of valleys that angle away from high areas toward adjacent lowlands. Each valley has been carved by water, and when viewed in cross section, has a "V-shaped" appearance. During alpine glaciation snow, firn, and ice accumulate in the upper reaches of these valleys and create steep-sided, bowl-shaped depressions called cirques. Ice flows out of a cirque and down a stream valley, eroding both soil (overburden) and bedrock. This erosion forms a U-shaped valley that is smoother and less sinuous than the original stream valley.

A series of small depressions may be carved out of solid rock by the valley ice. These depressions become filled with water when the ice melts and are known as tarns. Tarns are usually present in cirque basins and may occur at various places along associated valleys, in which case they have been called paternoster lakes because they resemble a string of pearls. Where three or more cirques from below the summit area of a mountain, ice plucking and abrasion may form a triangular peak known as a horn. The near vertical headwall of each cirque forms some of the steep upper slopes of the triangular peak. Above the headwall intense mechanical weathering results in very irregular topography.

Adjacent cirques and U-shaped valleys are often separated by a knife-like ridge known as an arete. Isolated low area or passes along the trend of an arete are known as cols and allow a hiker access to adjacent valleys.

Where a major valley (trunk stream) is intersected by a smaller valley (tributary stream), valley ice will carve a U-shaped pattern in both valleys. However, the thicker ice in the trunk stream will erode its valley deeper than the ice in the tributary stream forming a hanging valley.

The above mentioned features are illustrated in the following diagram.

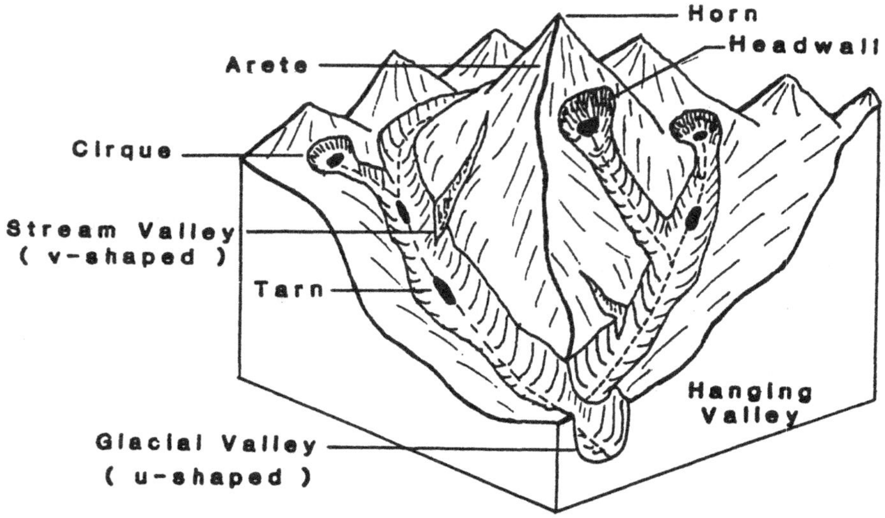

DEPOSITIONAL FEATURES OF ALPINE GLACIATION

When either valley ice or an ice sheet advance over land, both soil and pieces of bedrock are mixed with the ice. Thus, the ice contains variable amounts of rock and mineral fragments as it flows over the Earth's surface. Melting of the ice releases this debris which may be deposited directly upon the ground as unsorted and unstratified glacial till, or it may be transported by melt water for some distance and then be deposited on the ground as sorted and stratified glacial outwash.

Unsorted and unstratified material is laid down directly beneath the melting ice as ground moraine. Other types of morainal deposits include terminal, recessional, medial and lateral moraines. Terminal moraines (type of end moraine) appear as elongate mounds of glacial till that mark the farthest advance of the ice, whereas recessional moraines (another type of end moraine) mark places where the melting ice stood stationary for a short time during an otherwise retreating phase. Lateral moraines and medial moraines are associated with valley ice and mark the edges and more medial positions of the ice, respectively. They show up on aerial photographs as gray bands on the surface of the ice or as elongate mounds of glacial till on the valley floor.

Out in front of the melting glacier, quantities of glacial material are deposited by the meltwater that take the form of an outwash plain. Depressions in this outwash plain, which have resulted from the melting of buried blocks of ice, are known as kettles (kettle lakes if they are filled with water). Larger glacial lakes may also dot this outwash plain. Streams or rivers flowing into these lakes may from small deltas. Some of these features are illustrated below.

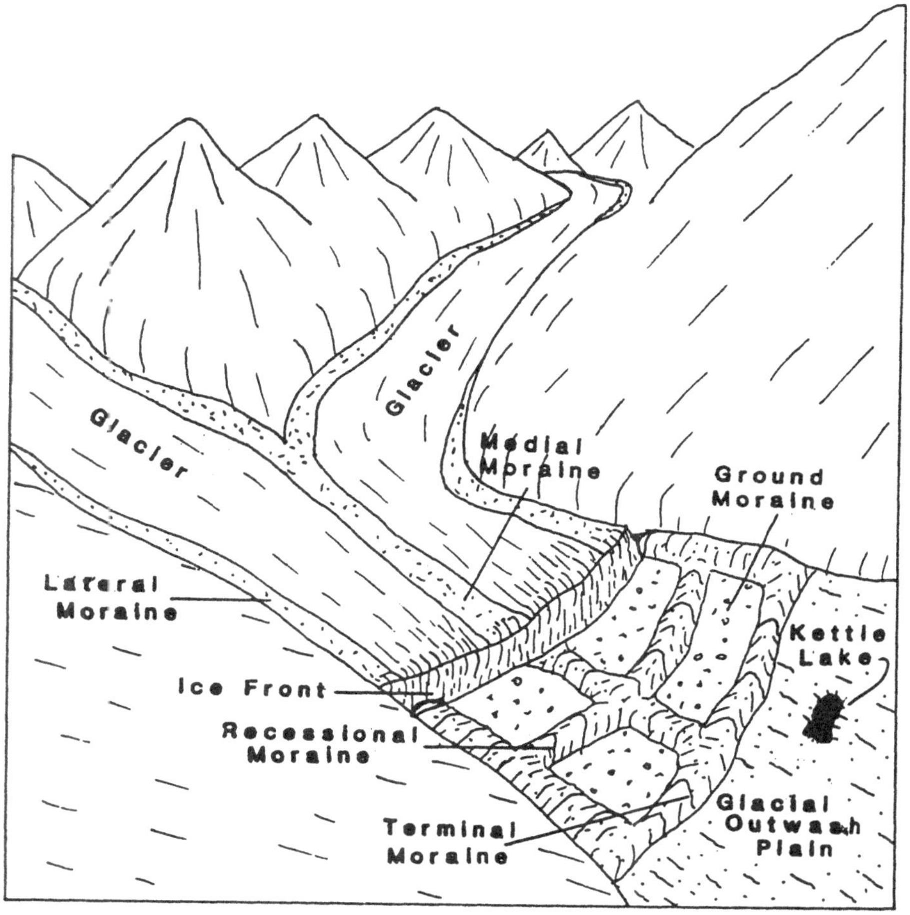

EXERCISE A

The Holy Cross, Colorado Quadrangle map covers a portion of west-central Colorado near the old mining town of Leadville. This area underwent extensive alpine glaciation during the Last Great Ice Age, but shows no glacial activity today. Answer the following questions by referring to the Holy Cross Quadrangle map:

1. Examine the topography in the vicinity of Homestake Peak located just west of the center of the map.

   a. What glacial term best describes the steep basin like features surrounding Homestake Peak?____

   b. What glacial term best describes Homestake Peak? __________________.

   c. Locate Lost Lakes northwest of Homestake Peak. What glacial term is applied to these lakes? ____
   and how did they form? ____
   ____.

   d. What glacial feature is responsible for the formation of Turquoise Lake?
   ____
   and what type of glacial material is this feature composed of? __________
   ____.

   e. Draw a topographic profile using only index contours across Lake Fork Creek from Bear Lake south to Sugar Loaf Mountain using elevation lines on the next page.

11,500 ____

11,250 ____

11,000 ____

10,750 ____

10,500 ____

10,250 ____

10,000 ____

9,750 ____

<u>NORTH</u> <u>SOUTH</u>

f. Does the shape of this profile suggest that the valley was occupied by ice at one time? ________ and how thick was the valley ice? ______________.

2. Examine the area around Turquoise Lake and the valley leading into the lake from the West.

a. How would you describe the topography near the northern, eastern, and southern shore of this lake? ________________________________.

b. The valley leading into Turquoise Lake starts near the western edge of the map. Name two glacial features developed in this area.

____.

c. Give evidence that valley ice flowed east to the vicinity of Turquoise Lake.

____.

3. Locate Galena Mountain near the center of the SW1/4 of this map.

a. How did this mountain become such a prominent peak? ______________

____.

b. What glacial term best describes the east-west ridge extending away from Galena Mountain? ______________________________. If you were to cross this ridge, what glacial feature would you use to make your traverse easier? ______________

4. Why are there so few lakes and ponds in the NE 1/4 of this map as compared to the NW 1/4 of this map? ______________________________

EXERCISE B

Examine the aerial photographic pairs under the stereoscopes located on laboratory tables. The topography on these photographs will appear in three dimensions and is the result of erosion tied to alpine glaciation. Name the geologic features on the photographs that are lettered A through J, in the space provided below:

A. ______________________

B. ______________________

C. ______________________

D. ______________________

E. ______________________

F. ______________________

G. ______________________

H. ______________________

I. ______________________

J. ______________________

EXERCISE C

Examine the large color photograph of the Teton Mountains of western Wyoming that is hanging on the laboratory wall. A large normal fault has lowered the land upon which the Snake River presently flows. Alpine glaciation has modified the Precambrian metamorphic and igneous rock exposed in the mountains. Answer the following questions with respect to this photograph:

1. Name 4 glacial features associated with the mountains:
   a. ______________________ b. ______________________
   c. ______________________ d. ______________________

2. Name the type of material upon which the Snake River flows:
   ____.

3. What name should be given to the elevated flat surface next to the Snake River?
   ____.

FEATURES ASSOCIATED WITH CONTINENTAL GLACIERS

Prior to the advance of continental ice sheets, topographic features were similar to that which we see today throughout the southern part of this country. Hills and mountains were somewhat rounded and most smaller stream and river valleys showed a typical "V" shaped pattern in cross section. Larger river valleys, as for example the Mississippi River valley, were very wide and had a well developed flood plain upon which the river meandered.

As climatic conditions changed and became more favorable for the formation of snow, firn and ice, ice sheets begin to spread in all directions away from centers of accumulation. In North America centers of ice accumulation formed in both Greenland and in the Hudson Bay area of Canada. This ice began its slow movement toward the south perhaps 3.0 million years ago (late Cenozoic Era), eventually covering nearly all of Canada and much of the northern part of the United States of America. Although these ice sheets advanced and retreated many times, the last episode of extensive glaciation in both North America and Europe reached its peak only about 18,000 years ago. These ice sheets have been retreating ever since that time because of a gradual warming trend during the last 18,000 years.

Both erosional and depositional features may form in response to continental glaciation. Erosion associated with ice sheets may involve removal of both bedrock and soil. These ice sheets become so thick that ice covers nearly all landforms except for a few prominent mountain peaks. Thus most mountains are overrun by the advancing ice, undergo erosion, and become more rounded. Clearly all lower areas, particularly stream and river valleys, are modified by the ice. Modifications include both deepening and straightening of preexisting valleys, and formation of "U" shaped valleys from original "V" shaped valleys.

Glacial ice contains appreciable amounts of rock and mineral material picked up by the ice as it advances over soil and bedrock. When the ice sheet melts in response to a climate change, this rock and mineral material is released and is deposited to form some kind of glacial deposit called glacial moraine. After the ice sheet has advanced to its farthest point, the ice sheet usually becomes stagnant for a time as the climate gradually shifts from cold to somewhat warmer. During this period of stagnation melting results in an accumulation of glacial debris in the form of an irregular ridge that parallels the leading edge of the glacier. This ridge of glacier till is called a terminal moraine, and is illustrated below.

TERMIAL MORAINE
(Cross Section)

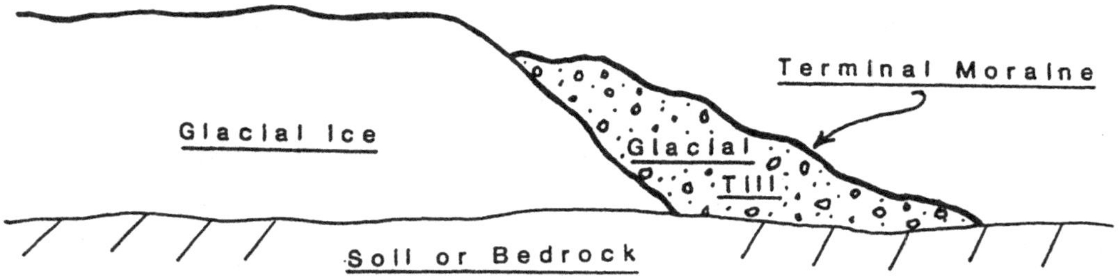

In front of a terminal moraine glacial debris is deposited in the form of a thin apron of sediment by glacial meltwater. This feature is referred to as a glacial outwash plain and consists of sorted and somewhat stratified glacial outwash that shows characteristics of water deposited sediment. During the formation of the terminal moraine and/or the glacial outwash plain, large blocks of ice break away from the melting glacier and become buried within either feature. When these blocks of ice melt a depression results that may or may not become filled with water. Individual depressions are called kettles. Water-filled kettles are called kettle lakes. These features are illustrated below.

KETTLES AND KETTLE LAKES
(Cross Section)

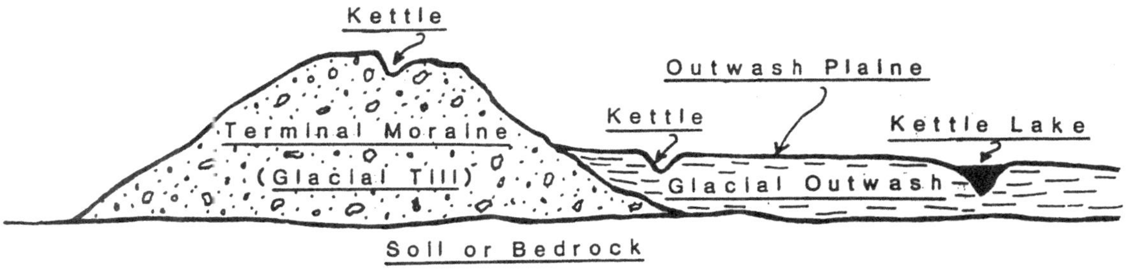

Continued melting of the ice results in a further retreat of the ice sheet, leaving the terminal moraine and the outwash plain far out in front of the glacier. A slow, but continuous retreat of the glacier results in deposition of an uneven blanket of glacial debris behind the terminal moraine. This blanket of glacial till, known as ground moraine, is neither sorted nor stratified because it has not been transported by water.

A ground moraine fills preexisting valleys and even covers low hills to produce an uneven surface after the ice has melted. New streams and rivers eventually begin to flow over this uneven surface. A deranged drainage pattern results because the streams and rivers tend to follow a course unrelated to any preglacial stream pattern. These streams and rivers tend to flow into and out of glacial lakes and cross many swampy areas. The sketch map below illustrates a typical deranged drainage pattern.

DERANGED DRAINAGE
(Map View)

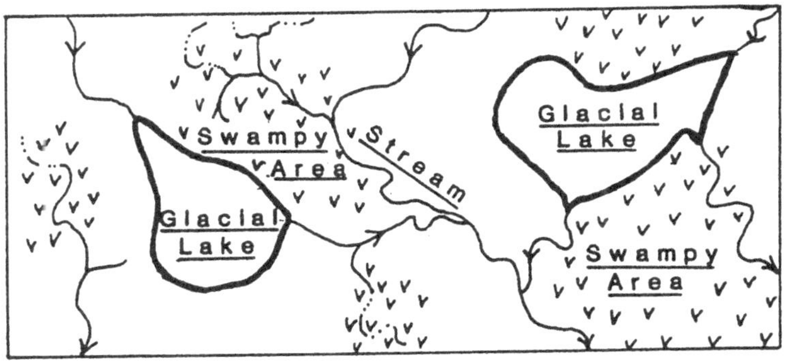

Continental glaciers have advanced and retreated many times during the Last Great Ice Age. A retreating ice sheet deposits glacial till over vast areas in the form of ground moraine. Decreasing temperatures combined with increased precipitation causes a retreating glacier to become an advancing glacier once again. As an ice sheet readvances over previously deposited glacial till (ground moraine), the ice may mound the glacial till into small hills with an elliptical shape called drumlins. The long axis of each drumlin indicates the general direction of ice movement and the steep side of the drumlin faces the direction from which the ice came. The diagrams below show these relationships in map view.

DRUMLINS

Map View

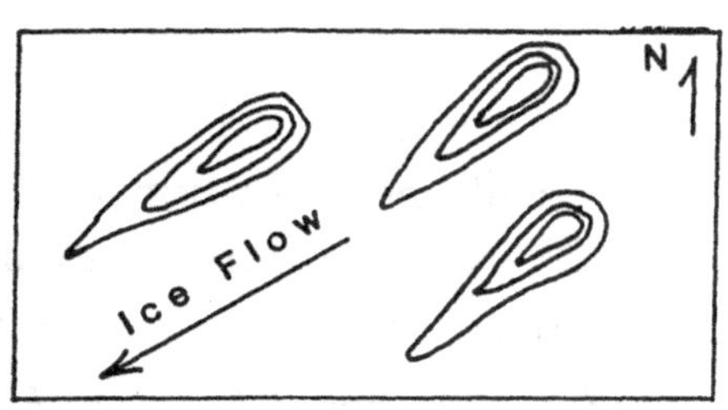

Cross Section

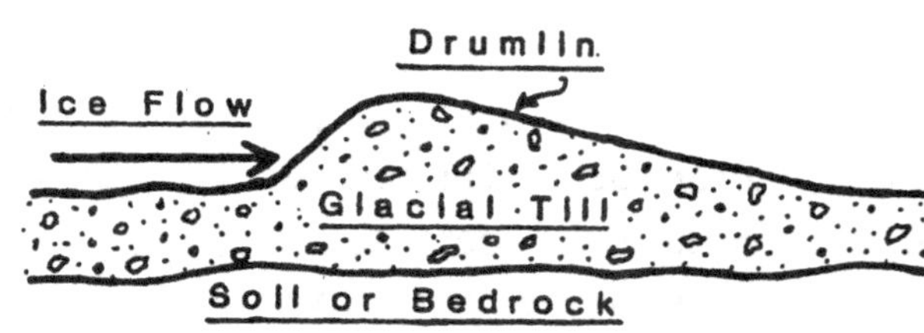

EXERCISE A

The Whitewater, Wisconsin Quadrangle covers a portion of southeastern Wisconsin that was covered by continental glaciers a number of times during the Last Great Ice Age. Numerous topographic features on this map demonstrate previous glacial activity. Answer the following questions by referring to the Whitewater Quadrangle.

1. Examine the topographic feature extending from the Southwest corner of the map to Sec. 17, T4N, R16E.

   a. Describe this feature in general terms ______________________________.

   b. What name should be applied to the material comprising this feature?
   ____.

   c. What name should be applied to the entire topographic feature? __________
   ____.

   d. Examine the relatively flat surface extending toward the Southeast from the feature named in question 1c above.

      1) What material comprises this feature? __________________________.

      2) What name should be applied to this entire feature? ____________.

      3) What name should be applied to the feature located near the center of Sec. 31, T 4N, R 16E? ________________. How did this feature form? ____.

   e. What is the approximate elevation of the bottom of the feature located near the center of Sec. 34, T4N, R15E? __________________________.

   f. Examine the drainage between Kettle Moraine State Forest and Whitewater.

      1) Why are there so many swamps in the area? ____________________.

      2) What name should be assigned to the drainage pattern of this area?
      ____.

g. Examine the many elliptical hills in the northwestern part of the quadrangle.

   1) What type of material comprises these hills? ____________________.

   2) What name should be applied to these hills? ____________________.

   3) What direction did the ice sheet move in this area? ______________.

CHAPTER 12

FOLDING AND GEOLOGIC HISTORY

by

Ronald H. Konig

INTRODUCTION

In lecture you were introduced to the concept of rock deformation. Most rock deformation is tied closely to the Theory of Plate Tectonics, which requires the lithosphere to be broken into large segments of rock known as plates. Each plate is free to move with respect to its nearest neighbor. Where plates approach each other and collide a Converging Plate Boundary and compressional forces result. Plates that move away from each other produce a Diverging Plate Boundary and tensional forces develop. Other plates simply slide past each other to form a Transform Plate Boundary and shearing forces. It is near these plate boundaries that compressional, tensional and shearing forces cause rock deformation in the form of different types of folds and faults. An appreciation of folding and faulting is not only important in the search for our dwindling natural resources, but is important in understanding certain environmental problems and solutions to those problems.

STRUCTURAL FEATURES IN ROCK

Igneous, sedimentary and metamorphic rock may undergo both folding and faulting. Folds may be so large that you would have to walk many miles to go from one side of the fold to the other side, or they may be so small that a microscope may be needed to see individual folds. Faults may be either large or small, but regardless of size always represent a fracture or physical plane of weakness in the rock along which there has been recognizable movement or sliding.

FOLDING

Anticlines and Synclines Although anticlines and synclines may occur in any kind of rock, they are most easily seen in sedimentary rock. It is the layering in sedimentary rock that makes it easier to see the bending associated with folding and the displacement or off-setting of beds associated with faulting.

Sedimentary rock always starts out as nearly horizontal sedimentary layers formed in a sedimentary basin. Lithification or consolidation of this sediment forms sedimentary rock with nearly horizontal layers or beds (figure 12-1).

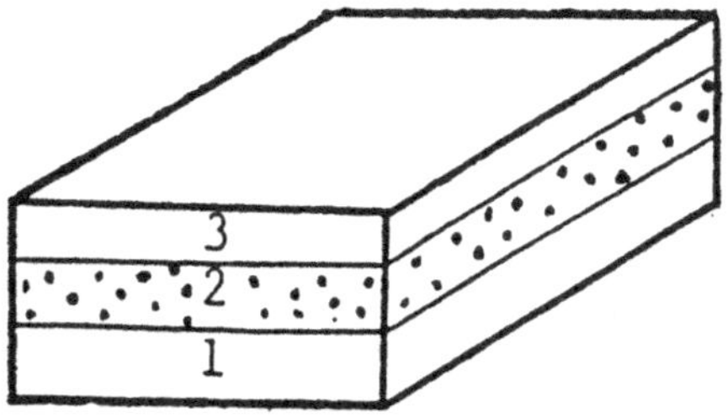

Figure 12-1. Block diagram showing horizontal sedimentary rocks, where layer 1 is the oldest layer and 5 is the youngest layer.

Compressional forces formed near a converging plate boundary may bend rock to form anticlines or synclines. An Anticline is an arch-like structure with older beds near its center that are flanked by progressively younger rocks. In contrast, a Syncline is a basin-like structure with younger beds near its center that are flanked by progressively older beds. These folds may be divided into two nearly equal parts by an imaginary Axial Surface that will intersect each bed surface to form the Hinge Line of the fold. Figure 12-2 shows an adjacent anticline and syncline in three dimensions that illustrates these features.

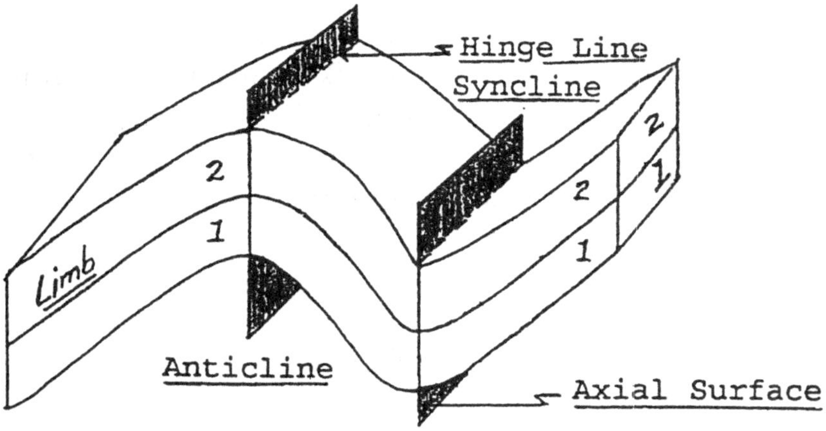

Figure 12-2. Three dimensional view of an adjacent anticline and syncline showing location of axial surfaces and hinge lines. Layer or bed 1 is older than bed 2.

Because the surface of the earth is a surface of erosion, the upper portion of these folds have usually been removed, producing a geologic map pattern similar to that shown in figure 12-3. The orientation of individual layers of rock or rock formations is shown on geologic maps by means of strike and dip symbols. The Strike of a rock formation is its trend with respect to True North, whereas Dip refers to the inclination of rock units away from the horizontal in a direction perpendicular to the strike. Figure 12-3 shows an adjacent anticline and syncline whose shape is indicated by a series of strike and dip symbols on the map surface. This figure also clearly shows younger rocks (Tr) exposed near the center of the syncline and older rocks (M) exposed near the center of the adjacent anticline. The front cross section of figure 12-3 shows the true shape of the these folds.

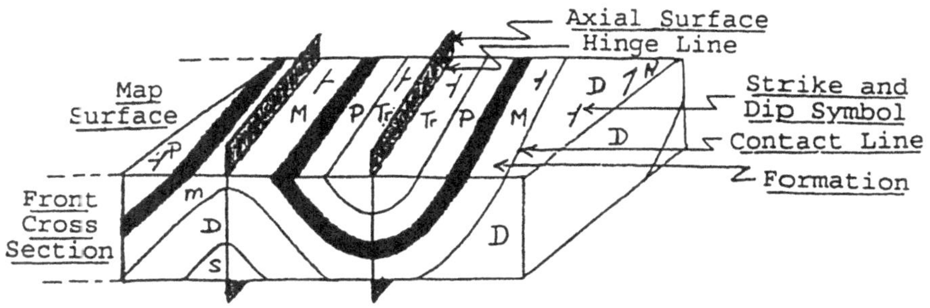

Figure 12-3. Block diagram of an anticline and syncline showing axial surfaces, (dashed lines), hinge lines, strike and dip symbols and relative ages of rock units expressed by symbols taken from the geologic time chart.

Two-dimensional and three-dimensional diagrams of both undeformed and deformed rock will be accompanied by a geologic legend or explanation (table 12-1). A Geologic Legend explains the meaning of all lines, numbers, letters and symbols used in the diagram and often indicates the relative age of the rock units (formations) using either numbers or approved symbols form the geologic time chart. All geologic time, as noted in an earlier laboratory exercise has been divided into packages of time known as Eras, Periods and Epochs that define the Geologic Time Chart. A simplified version of the time chart used by us is given in table 12-2.

GEOLOGIC LEGEND	
Rock Unit Symbols	**Other Symbols**
P IP M D S — See geologic time chart below for these abbreviations	Strike & Dip Symbol Formational Contact Anticline Syncline

Table 12-1. Sample geologic legend for a geologic map.

Era	Period	Symbol	Millions of Years Ago
Cenozoic	Quaternary Tertiary	Q T	 66.4
Mesozoic	Cretaceous Jurassic Triassic	K J T_R	66-144 144-208 208-245 245
Paleozoic	Permian Pennsylvanian Mississippian Devonian Silurian Ordovician Cambrian	P IP M D S O €	245-286 286-320 320-360 360-408 408-438 438-505 505-545 545
Proterozoic Archeozoic	Precambrian Time P€		

Table 12-2. Geologic time chart

Anticlines and synclines not only occur together, but are often inclined away from the horizontal. Folds inclined away from the horizontal are said to Plunge. Plunge produces a very characteristic "U"-shaped rock pattern on the earths surface. Figure 12-4a is a three dimensional illustration of an anticline that is plunging due north. Erosion will remove the upper part of the anticline and produce the characteristic "U"-shaped pattern illustrated on the map surface of figure 12-4b. The direction of plunge may be determined by examining map patterns at the earths surface and by determining the dip direction of beds along the axial trace of the fold.

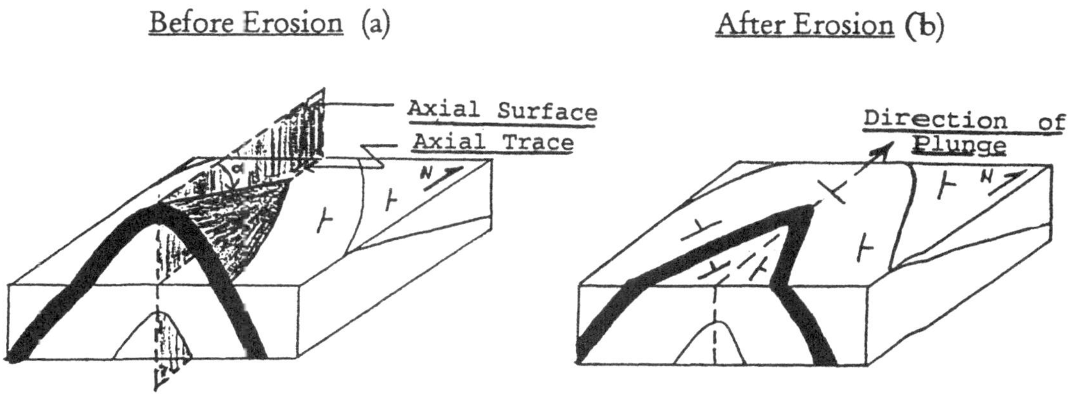

Figure 12-4. Three dimensional diagrams of a north plunging anticline as seen before erosion and after erosion.

Adjacent anticlines and synclines always undergo erosion at the earths surface. Geologic maps show the shape of these folds by means of contact lines, axial traces, and strike and dip symbols. Figure 12-5 illustrates the appearance of an adjacent anticline and syncline in both map view and in cross section after erosion has taken place.

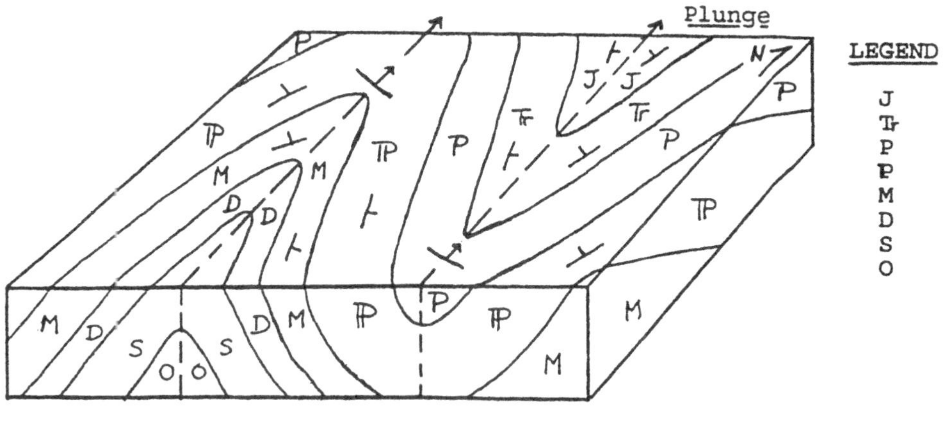

Figure 12-5. Three dimensional diagram showing north plunging folds in map view (top surface) and in front and side cross sections.

Rules for Interpreting Folds: A series of simple rules allows the Geologist to identify individual folds and to interpret the geologic history of the area. In each of the following diagrams lower numbers are assigned to older beds and higher numbers to younger beds.

Rule #1: In undeformed sedimentary rock, the youngest rock formation will overlie older rock formations. (Principle of Superposition)

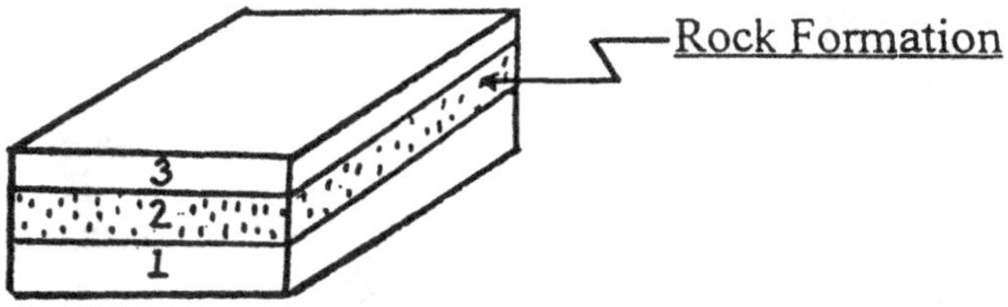

Rule #2: In deformed sedimentary rock, older rock formations will be inclined (dip) in the direction in which younger rock formations are located.

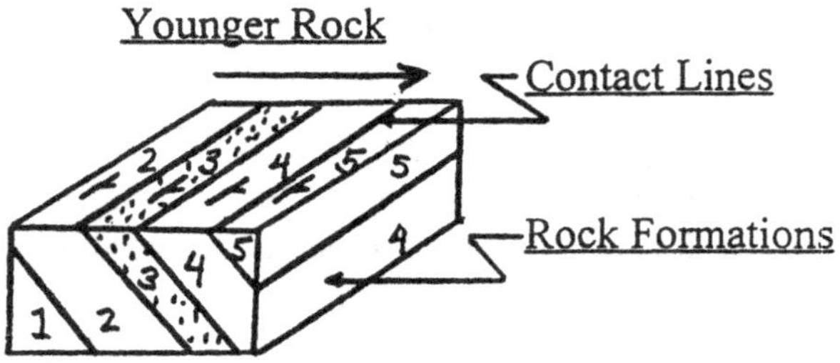

Rule #3: Older rocks are exposed in the central area of anticlines and younger rocks are exposed in the central area of a syncline.

Anticline

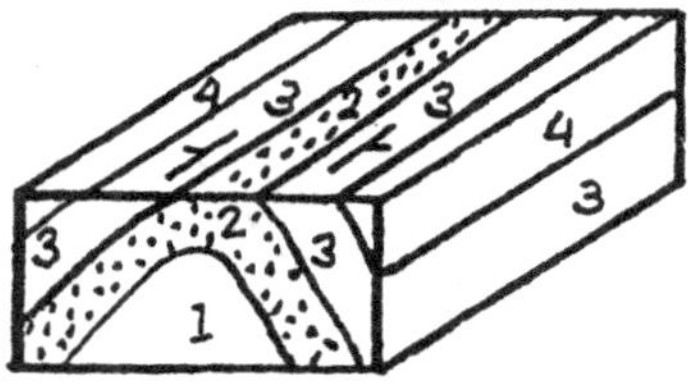

Syncline

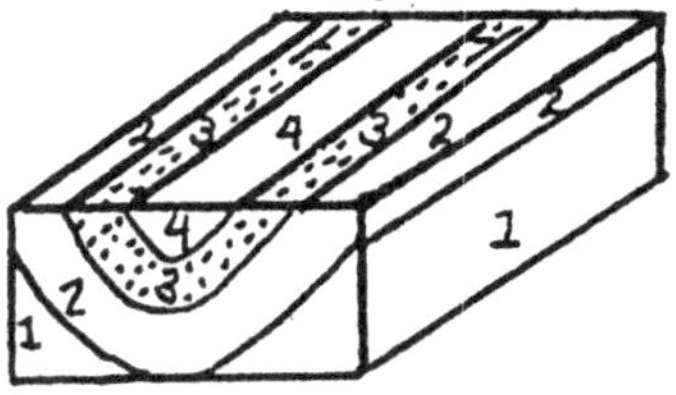

Rule #4: Non-plunging folds have straight contact lines that are parallel to each other, whereas plunging folds have contact lines that curve around its axial trace.

Non-Plunging Fold

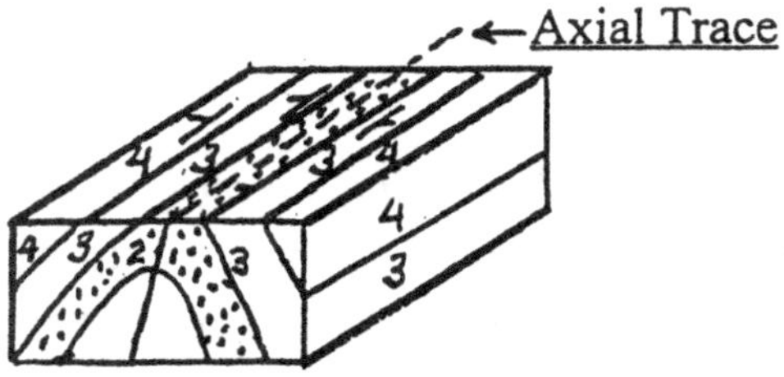

Plunging Anticline

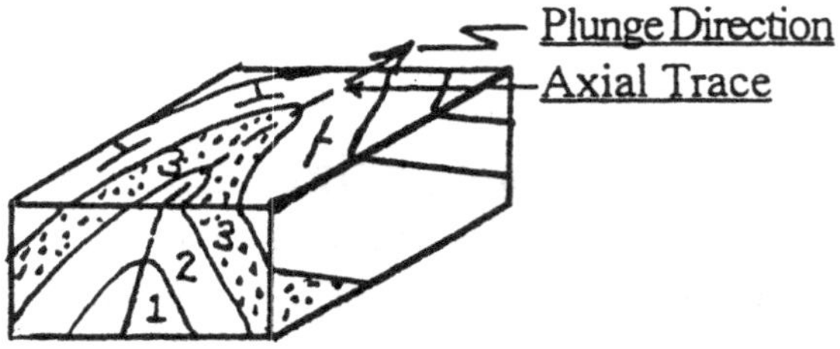

Exercises Involving Folded Sedimentary Rock: Each of the following sketches represents 1) a geologic map, 2) a geologic map with space for construction of a geologic cross section involving folded sedimentary rock, or 3) a geologic cross section. Answer the questions in the order in which they are given for best results

Fold Problems

1. Block Diagram

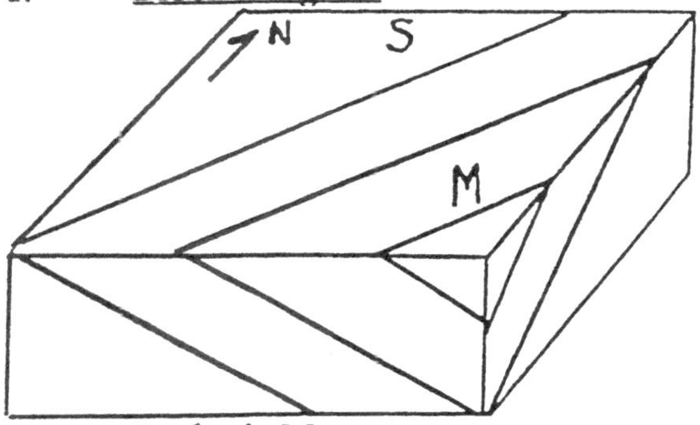

Questions

1) Label all formations with a symbol from Geologic Time Chart. Note "S" for Silurian Period.
2) Place a strike and dip symbol on each formation (Rule 2)
3) In what direction do the rocks strike? __________

2. Geologic Map

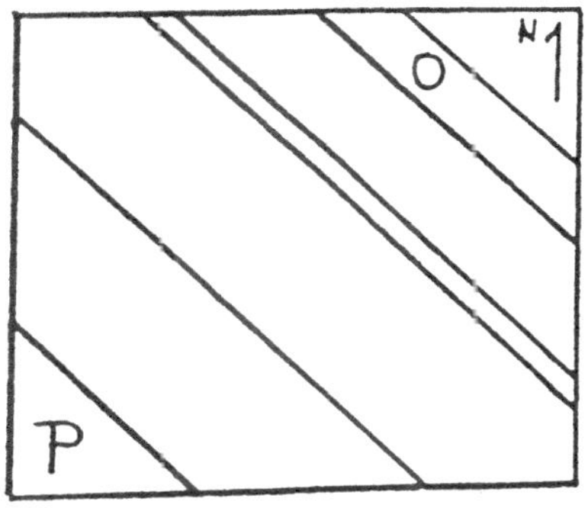

Legend

Questions

1) Label each formation with a symbol from the Geologic Time Chart.
2) Place a strike and dip symbol on each formation.
3) In what direction do the rocks dip? ___________
4) Establish a geologic legend showing age of rocks in space provided.

3. Geologic Map

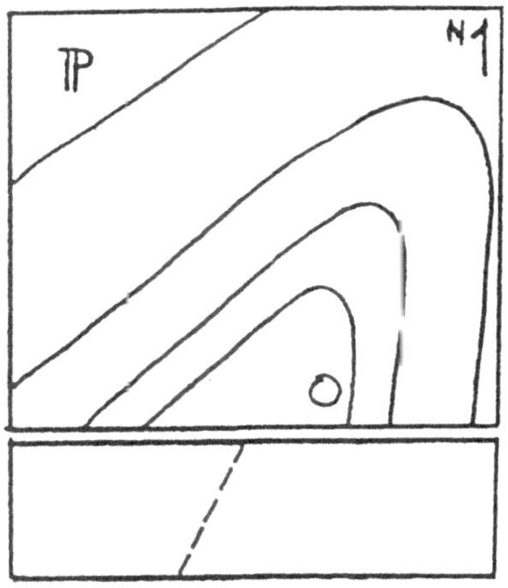

Cross Section

Questions

1) Place a strike and dip symbol on each formation shown.
2) Draw in the axial trace of the fold.
3) Label each formation with symbols form the Geologic Time Chart. Note "O" for Ordovician.
4) In what direction does this fold plunge? _____
5) Complete the cross section in the space provided.

4. Geologic Map

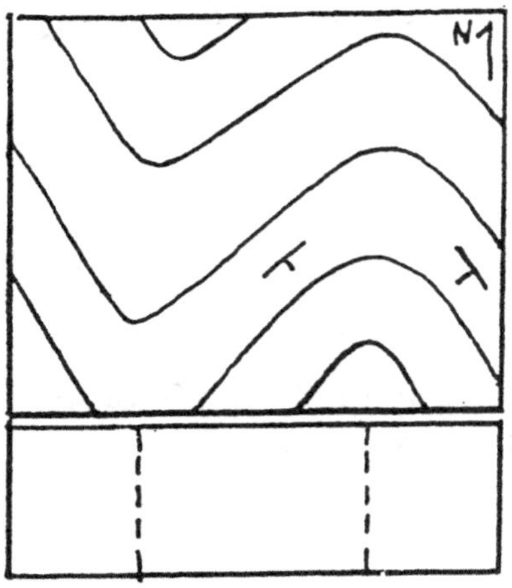

Legend

Questions

1) Label each formation with symbols from the Paleozoic Era of the Geologic Time Chart
2) Place strike and dip symbols on each formation.
3) Draw in the axial trace for each fold.
4) In what direction do these folds plunge? ________________________.
5) Complete the front cross section in the space provided.
6) Establish a geologic legend in the space provided.

5. Geologic Map

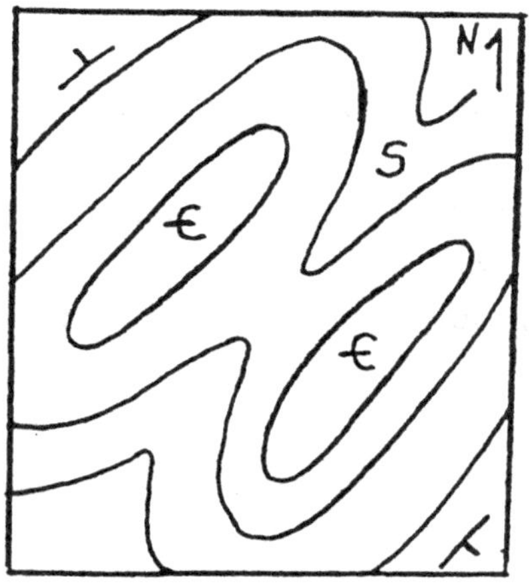

Legend

Questions

1) Label each formation with symbols from The Paleozoic Era.(Note strike and dip symbols)
2) Place strike and dip symbols on all formations.
3) Draw in the Three axial traces.
4) In what direction does the most southeasterly fold plunge? ____________________.
5) Label each fold shown as either an anticline or a syncline.
6) Establish a geologic legend in the space provided.

6. Outcrop Map

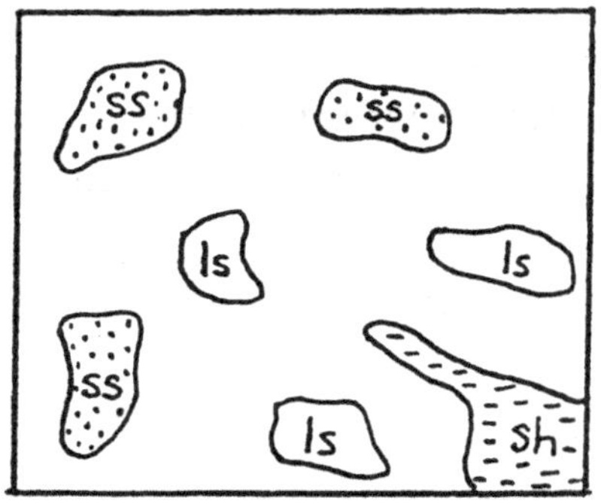

Legend

ss
ls
sh

Questions

1) Draw contact lines for each formation.
2) Draw in the axial trace of the fold.
3) Give the direction of plunge. ____________
4) Name the fold. ______________________

7. <u>Geologic Cross Section</u>

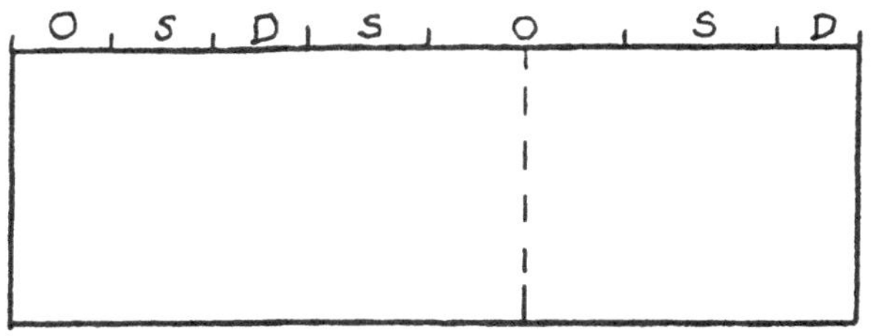

<u>Questions</u>

1) Using the age of the rock formations given along the top line, complete the geologic cross section.

8. Interpret the geologic history (Sequence of geologic events) from oldest to youngest by examining the geologic cross section below.

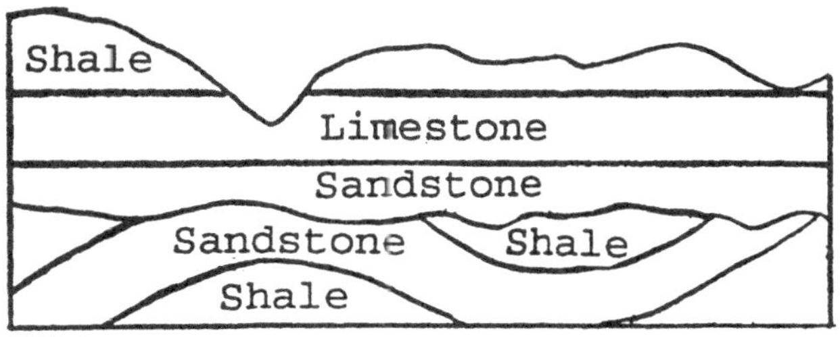

(Youngest Event)

10 ____________

9 <u>Uplift</u>

8 ____________

7 ____________

6 <u>Subsidence</u>

5 ____________

4 ____________

3 ____________

2 <u>Consolidation</u>

1 <u>Deposition</u>

Oldest Event

CHAPTER 13

FOLDING, FAULTING AND UNCONFORMITIES

by

RONALD H. KONIG

INTRODUCTION

In the previous laboratory exercise you were introduced to the shape of folding in both map view and in cross section, and learned how to interpret simple geologic history of folded rocks. In this laboratory exercise folding will be altered by both faulting and various kinds of unconformities. Proper understanding of folding, faulting, and unconformities allows the student to unravel the complex geologic history of deformed rock.

Mother Natures' Faults

Faults in the Crust of the earth are fractures or cracks in rock along which movement (sliding) has occurred parallel to the fracture surface. These faults result from application of tensional, compressional or shearing forces to rock. Tensional forces, in combination with gravity produces normal faults, compressional forces form Reverse Faults, and shearing forces result in Strike-Slip faults. All faults may be observed in map view and/or in cross section.

Normal Faults

A **Normal Fault** is any fault whose hanging wall has moved down with respect to its footwall. The terms hanging well and footwall were first used by miners who were excavating mineralized rock along a tunnel driven parallel to a fault that had been mineralized. The miner's feet would be on the footwall side of the fault and his head would be on the hanging wall side of the fault as he walked along the drift or tunnel (figure 13-1).

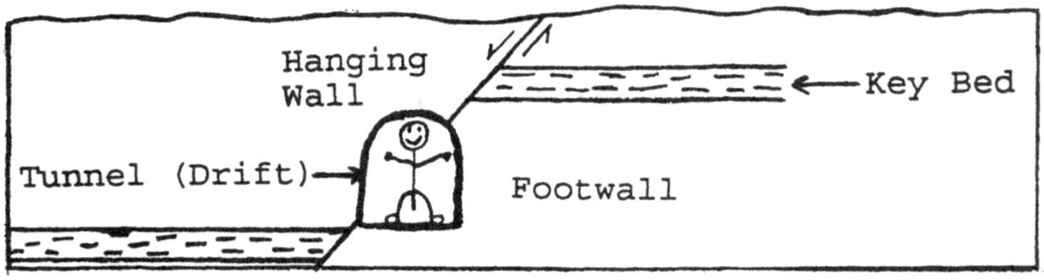

Figure 13-1. Cross sectional diagram of a drift (tunnel) that is cut by a normal fault showing the position of both the Hanging Wall and the Footwall.

Recognition of a normal fault requires locating a key bed. A **Key Bed** is any layer of rock that may be traced across the fault even though faulting may have displaced the bed many tens or even hundreds of feet. Layers of sedimentary rock often serve as key beds. Figure 13-2 is a block diagram that shows the effects of normal faulting associated with horizontal sedimentary rock and figure 13- 3 involves normal faulting and inclined (dipping) beds as they would appear both before and after erosion.

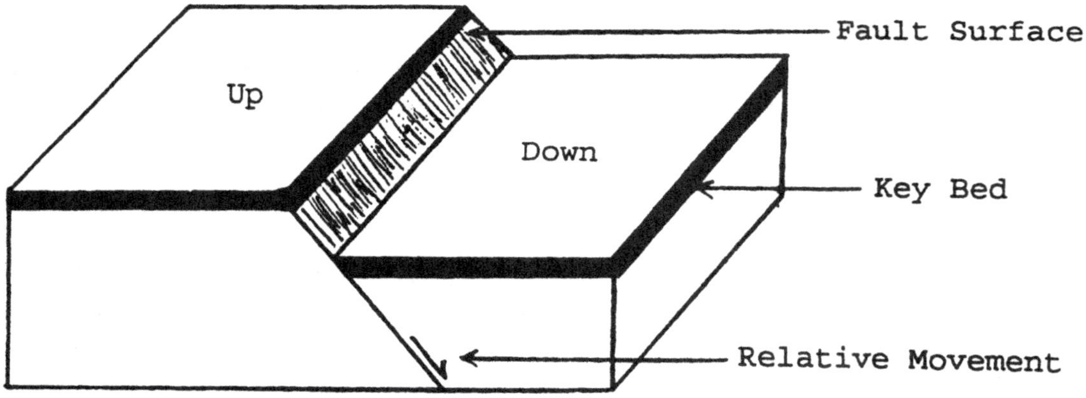

Figure 13-2. Block diagram showing horizontal sedimentary rocks that have been cut by a normal fault..

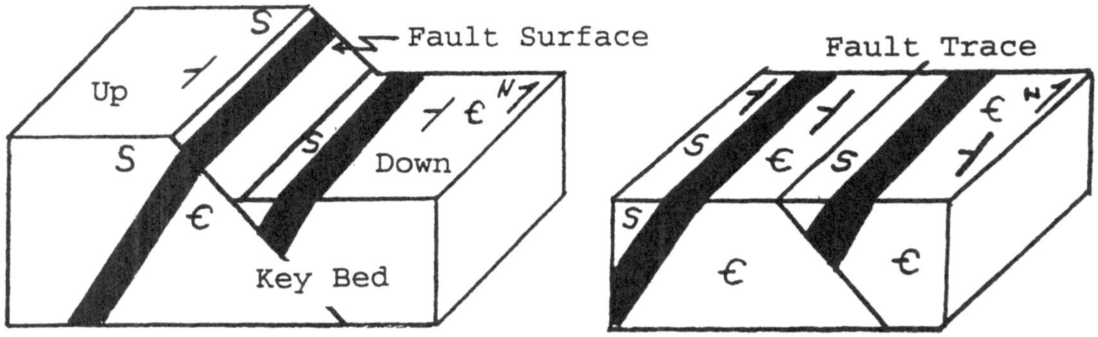

Figure 13-3. Block diagrams showing the effects of erosion on faulted sedimentary rock that is dipping 45 degrees.

Erosion tends to lower rock on the upthrown block of a fault to the level of rock on the downthrown side of the fault. After erosion, the new position of the key beds on either side of the fault trace may be seen. Faulting followed by erosion may also result in a repetition of rock units on either side of the fault trace as shown in figure 13-3. Clearly, normal faulting results in a stretching (elongation) of rock parallel to the earths surface as a result of tensional forces.

Reverse Faults

A **Reverse Fault** is any fault whose hanging wall moves down with respect to its footwall. Recognition of a reverse fault requires location of a key bed on both sides of a fault surface. Figures 13-4 are block diagrams of a reverse fault as it appears before and after erosion of the upthrown fault block.

Before Erosion

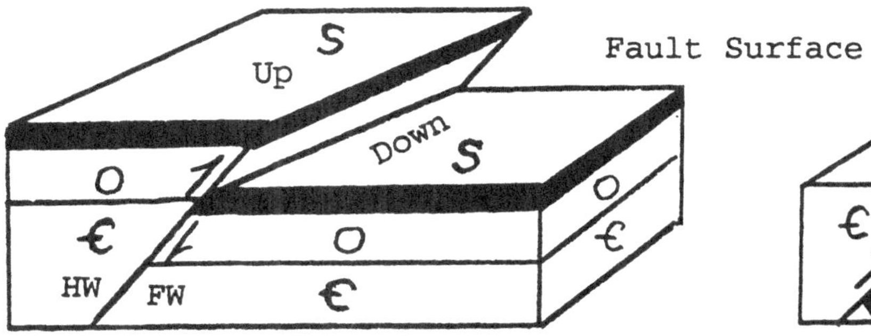

Figure 13-4a. Horizontal beds cut by a reverse fault.

After Erosion

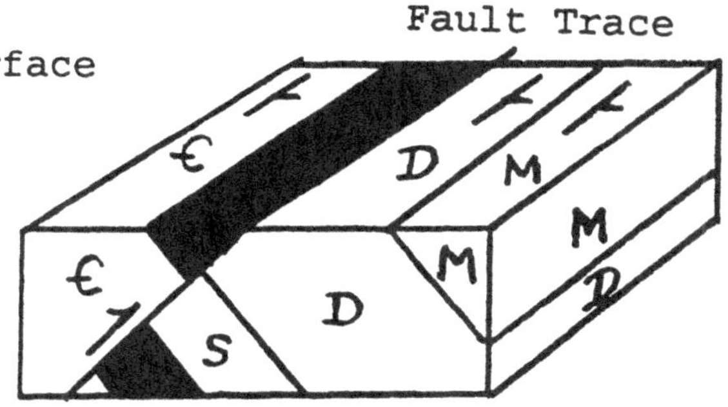

Figure 13-4b. Inclined beds but by a reverse fault showing omission of bed "S" in map view.

Reverse faulting of inclined beds that dip in a direction opposite to that of the fault will often result in an apparent omission of beds on a geologic map after erosion. Figure 13-4b shows the omission of Silurian age rocks (S) on the map surface as you move from left to right across the fault trace. Clearly, reverse faulting results in a shortening of rocks in the earths crust as a result of compressive forces.

Strike-Slip Faults

A **Strike-Slip Fault** is a fault, which often results from the application of shearing forces developed along transform plate boundaries, is a nearly vertical fracture where adjacent fault blocks move (slide) parallel to the strike or trend of the fault surface. A well known example of a strike-slip fault is the San Andreas fault in California. Any sudden movement along this fault produces an earthquake whose epicenter lies directly above where the movement occurred. Figure 13-5 shows the general appearance of a strike-slip fault in three dimensions.

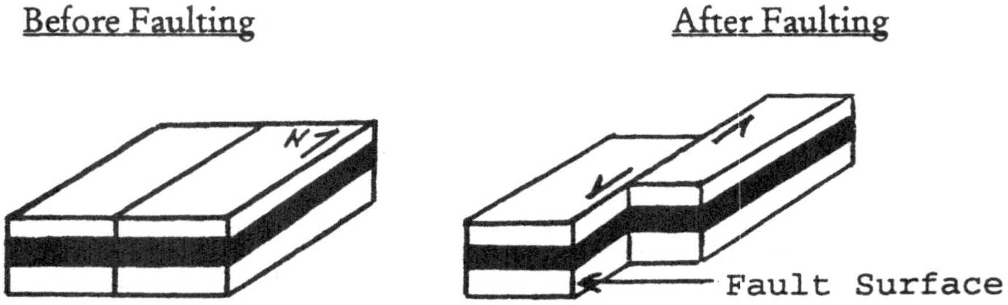

Figure 13-5. Block diagrams showing a typical strike-slip fault. Note that there is neither a hanging wall nor a footwall because the fault surface is vertical.

Folding and Faulting

In mountains ranges, like the Appalachian and Ouachita Mountains of North America, rocks have undergone both folding and faulting in response to forces resulting from lithospheric plate collision. Although the resulting picture may be very complex, it is usually possible to unravel what has happened to the rocks during mountain building. Geological study not only allows one to identify specific types of folds and faults in the area, but allows the geologic history of the area to be interpreted. Figure 13-6 is a three dimensional view of sedimentary rock that has been folded, faulted, and finally altered by erosion.

Before Erosion After Erosion

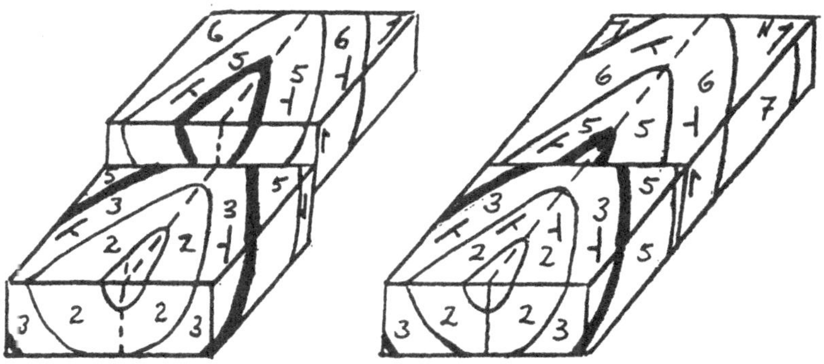

Figure 13-6. Three dimensional diagram showing a south plunging syncline cut by a south dipping normal fault both before and after erosion.

Note how erosion has resulted in migration of the beds in the general direction of their dip on the up thrown side of the fault.

Rules for Interpreting Geologic Structures: Anticlines, synclines, normal faults, reverse faults, and strike-slip faults are the most important geologic structures formed during mountain building along converging plate boundaries. A series of simple rules allows the Geologist to identify individual structures and to interpret the geologic history of the area. Rules used in evaluating dipping beds, anticlines and synclines were considered in the previous laboratory. Additional rules applicable to fault problems are listed below.

Rule #1: The upthrown side of a normal or reverse fault will be eroded (lowered) to the level of the down thrown side of the fault in time.

Normal Fault Reverse Fault

Rule #2: Contact lines and associated rock formations always shift in the direction of dip as the land surface is lowered by erosion.

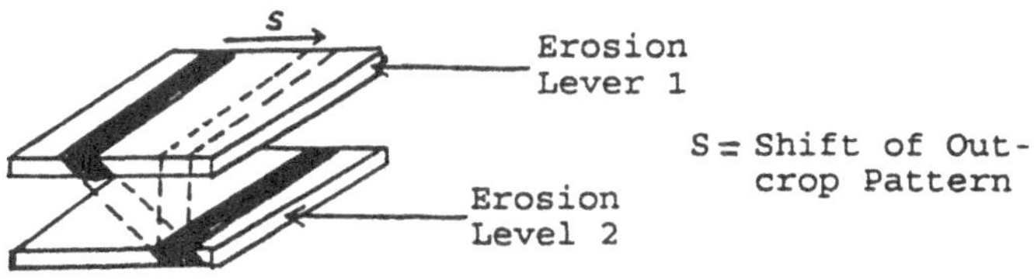

UNCONFORMITIES An unconformity is any natural surface within the earth that represents some kind of a break in earth history. The rock lying above this surface is always much younger than the rock below the surface. Unconformities may result from either erosion or lack of deposition. The three types of unconformities recognized are disconformities, angular unconformities and non-conformities.

A **Disconformity** usually results from erosion (removal) of sedimentary material followed by more deposition of sediment, to produce an erosional surface that is parallel to the younger rocks above and the older rocks below. Figure 13-7 shows how a Disconformity may result from erosion.

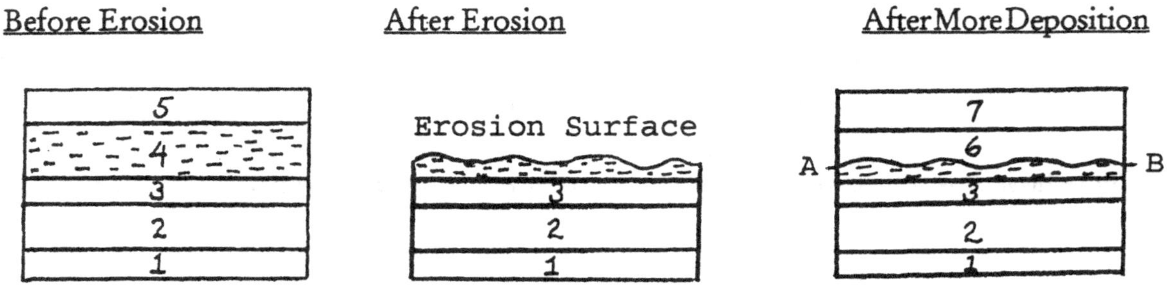

Figure 13-7. Cross sectional diagrams showing formation of a disconformity as a result of erosion.

An angular unconformity results from erosion of deformed sedimentary rock followed by deposition of sediment giving rise to younger nearly horizontal sedimentary rock. In figure 8, the sequence of events would be, from oldest to youngest, 1) deposition, 2) lithification to form sedimentary rock, 3) rock deformation, 4) uplift of the deformed rock, 5) erosion, 6) subsidence, and finally 7) renewed deposition to form younger sediment and eventually sedimentary rock.

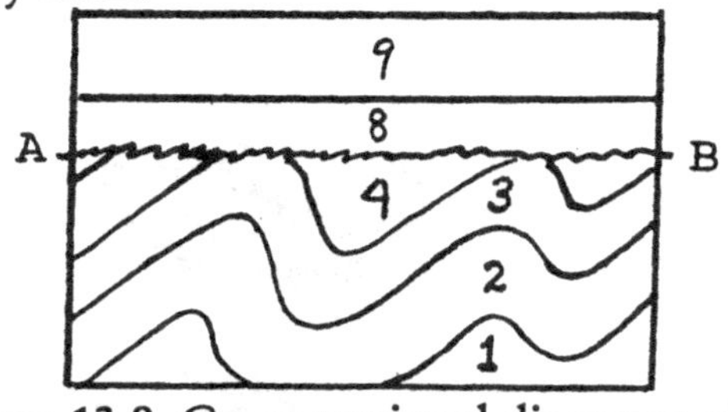

Geologic History

8) Lithification	4) Uplift
7) Deposition	3) Deformation
6) Subsidence	2) Lithification
5) Erosion	1) Deposition of 1-5

Figure 13-8. Cross sectional diagrams of an angular unconformity (AB.).

A Non-Conformity results form erosion of older igneous and/or metamorphic rock followed by deposition of younger sediment forming essentially horizontal sedimentary rock overlying igneous and/or metamorphic rock (figure 13-9). Both non-conformities and angular unconformities usually require considerable time to form because thousands of feet of younger rock must be eroded in order to expose the more deeply buried rock at the earths surface.

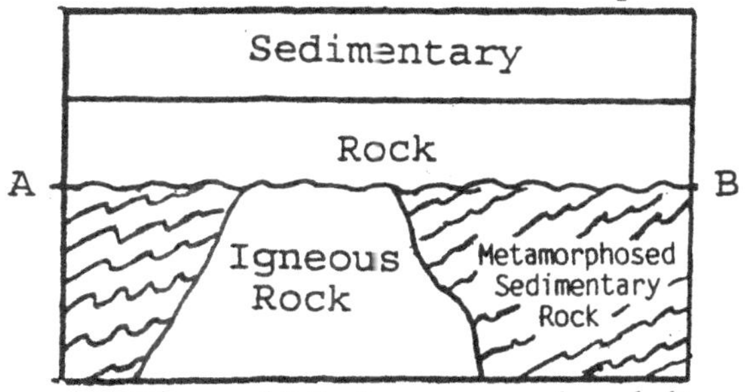

Geologic History

8) Lithification
7) Deposition
6) Subsidence
5) Erosion
4) Uplift
3) Igneous Intrusion Sediment
2) Deformation & Metamorphism
1) Deposition (oldest)

Figure 13-9. A cross sectional diagram showing a typical non-conformity (AB).

Exercises Involving Folding and Faulting Each of the following sketches is a geologic map showing a series of rock formations whose normal arrangement has been disturbed by either faulting or by folding and faulting. Answer the questions in the order in which they are given for best results.

1.

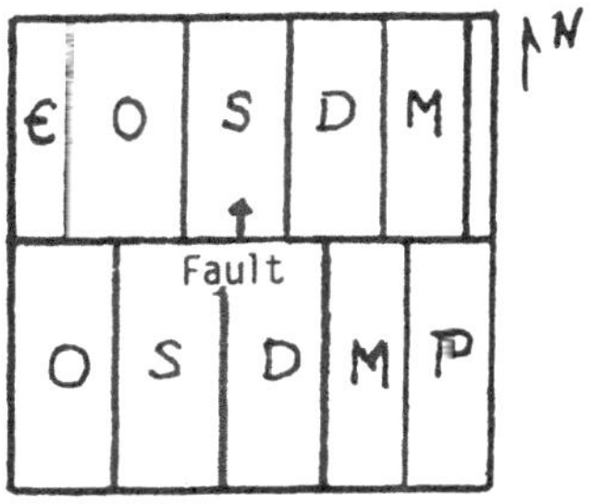

Legend

1) Compile a legend of the symbols on the map.
2) In which direction do the beds strike?________
3) In which direction do they dip? __________
4) Which is the upthrow side?_____________
5) Since the fault dips north, is it normal or reverse?______________

2.

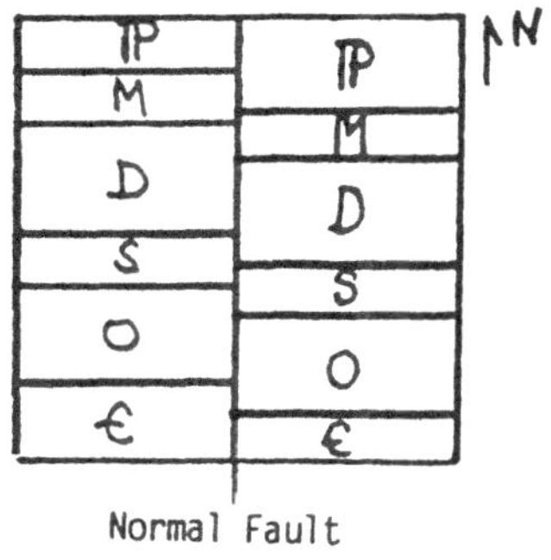

Legend

1) Compile a legend of all the formations on the map.
2) In which direction do the beds strike and dip?

____

3) Which is the upthrow side?_____________
4) In which direction does the fault plane dip?

____

3.

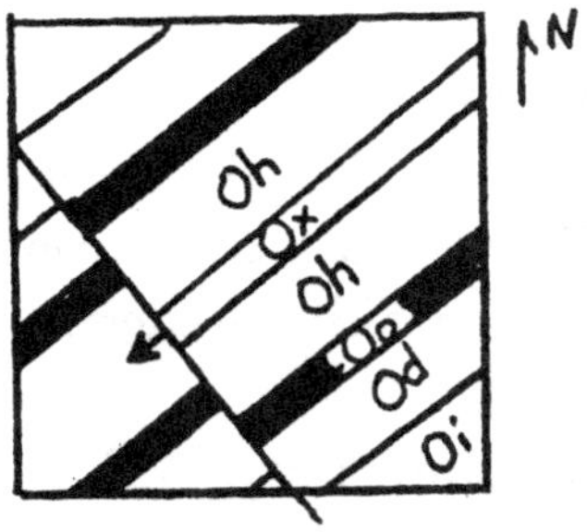

Reverse Fault

Legend

The abbreviations on the map are meaningless symbols. One bed has been shaded for identification.

1) Which is the upthrow side?____________
2) Has the fault cut across an anticline or a syncline? Reason?______________________________.
3) Finish labeling the formations on both sides of the fault.
4) Establish a geologic legend.

4.

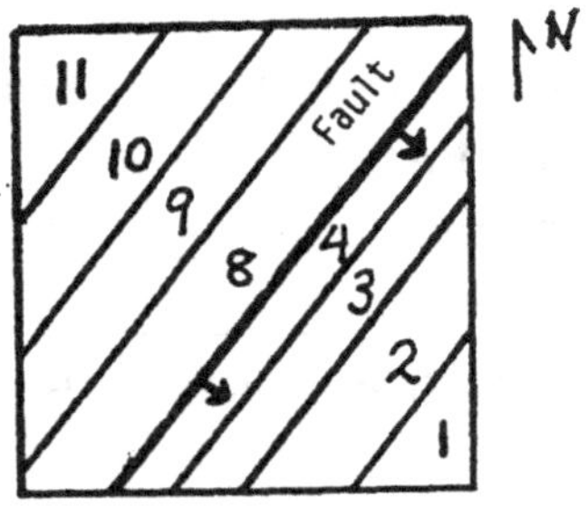

Legend

12
11
10
9
8
7
6
5
4
3
2
1

1) In which direction do the beds strike and dip?

______________________.

2) In which direction does the fault strike?
3) Which is the throw side?___________
4) If the fault dips southeast, is it normal or reverse?__________________

5. Construct a geologic cross section in the space below after examining the location and age of various rock formations shown along the topographic profile.

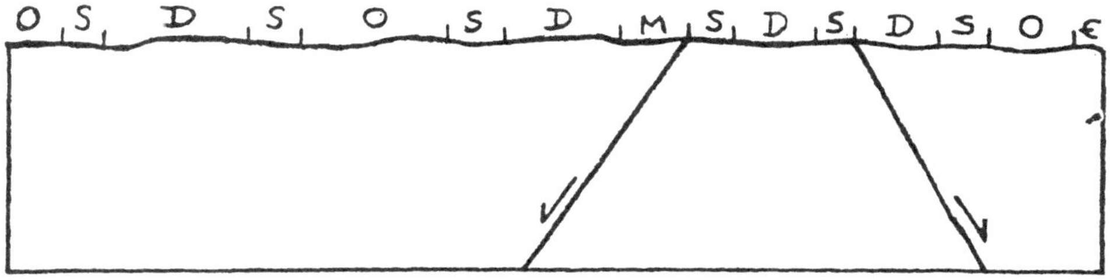

Exercises Involving Geologic History of an Area Each of the following cross sectional sketches shows the geology of a certain area. The geologic history of each area may be determined by examining the rock formations and associated geologic structures shown. Your assignment, if you should choose to except the challenge and earn full credit for this laboratory, is to interpret the geologic history of each area. This may be done by listing the geologic history geom. oldest to youngest in tabular form next to each diagram. Terms like deposition, erosion, uplift, folding, intrusion, it. should be used for each geologic history given.

1. Interpret the geologic history from this geologic cross section.

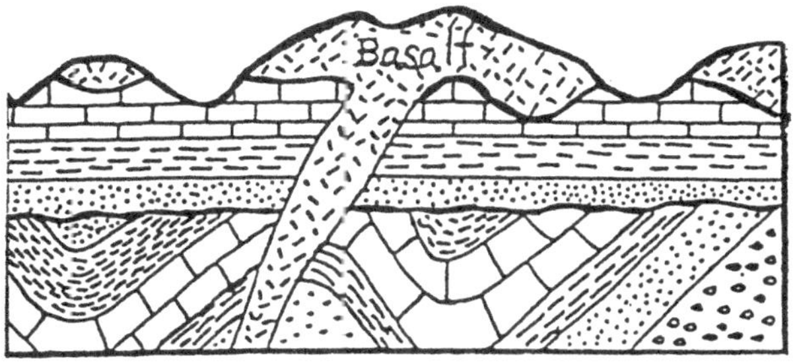

Geologic history (11 steps)

1. ______________ Youngest
2. ______________
3. ______________
4. Lithification
5. ______________
6. ______________
7. ______________
8. Uplift
9. ______________
10. ______________
11. Deposition Oldest

2. The accompanying generalized drawing shows the section exposed in the Grand Canyon of Colorado River in Arizona. Interpret the geologic history of the area.

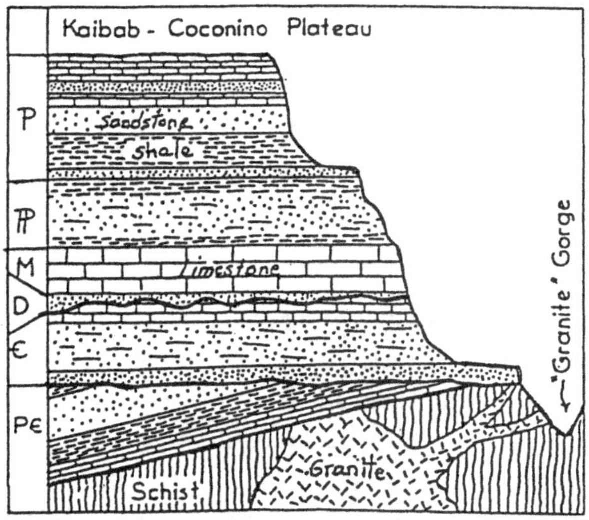

Geologic History (16 steps)

1. ______________ Youngest
2. ______________
3. ______________
4. Deposition
5. ______________
6. ______________
7. ______________
8. ______________
9. ______________
10. Subsidence
11. ______________
12. ______________
13. ______________
14. ______________
15. ______________
16. Deposition Oldest

APPENDIXES

APPENDIX A

THE RADON GAS PROBLEM

by

RONALD H. KONIG

INTRODUCTION

Some scientists think that a significant percentage of lung cancer not caused by smoking may be related to the presence of radon gas in our environment. Radon gas (radon-222) is present in rock, and has been part of our environment since the formation of the Earth about 4.5 billion years ago (bya). Anything that tends to concentrate this small amount of radon gas at or close to the earth's surface could pose a health risk for humans. In 1984 high concentrations of radon gas were detected in a home in southeastern Pennsylvania. Apparently some buildings can act as collectors for radon gas. For such concentrations to occur, radon must pass from rock in the crust of the earth into the overlying soil upon which a building has been constructed. Additionally, the basement or lowest part of the building must be relatively air tight. Thus some of the newer energy efficient buildings may retain greater concentrations of radon gas than older structures.

Weathering of any exposed rock gives rise to unconsolidated material called soil. Soil often serves as foundation material for buildings. If radon gas is present either below or within the soil it may enter a building through any opening in a foundation. Such openings include cracks in floors and walls, pore spaces in cement blocks, and sumps equipped with pumps.

CHEMICAL RELATIONSHIPS OF RADON-222

Chemists have established that radon-222 (a gas) has as its ultimate source the element uranium-238. Geologists have discovered that uranium-238 is associated with natural uranium-rich minerals like uraninite, pitchblende, carnotite, etc. These minerals are not distributed uniformly throughout the crust of the earth. Certain rocks like granite, gneiss, sandstone, black shale, etc. may have unusual concentrations of uranium-rich minerals. Uranium-238 is not a stable element and decays (breaks down) with time into new, lighter elements as shown in figure 1.

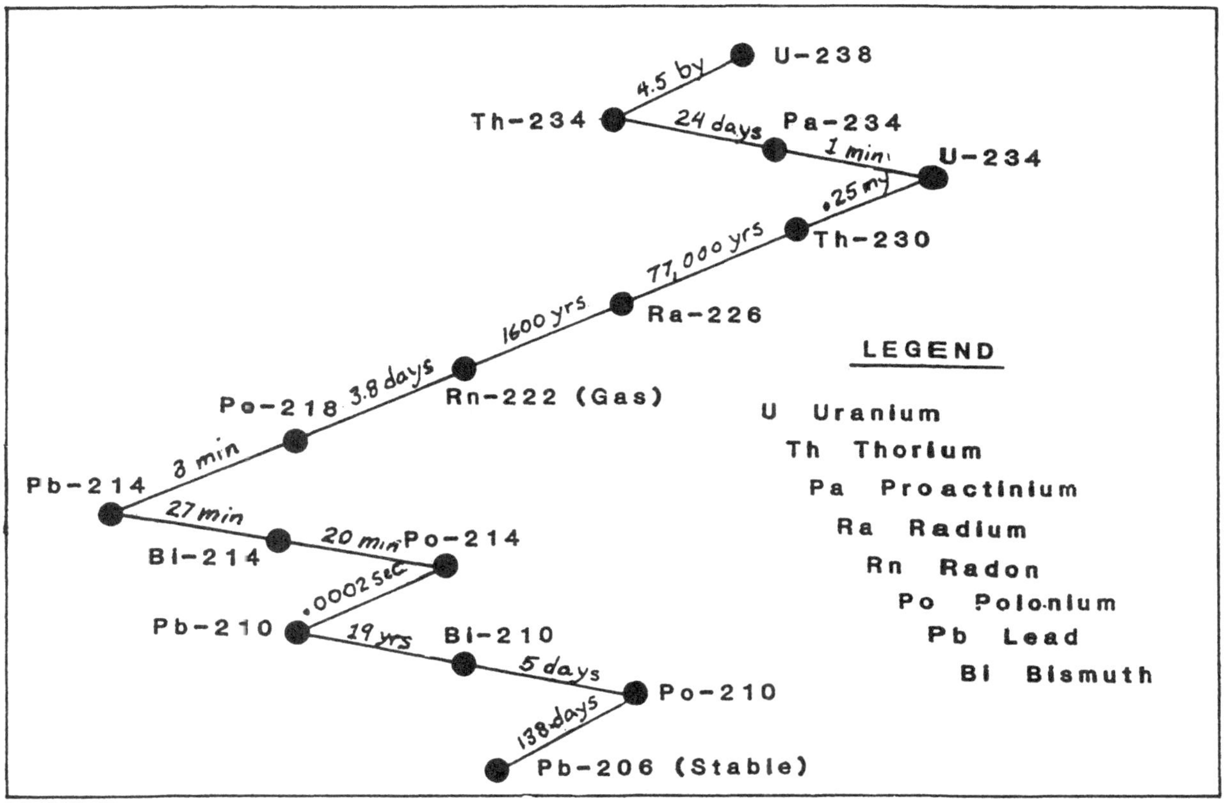

Figure 1. Schematic diagram showing how uranium-238 decays naturally to form new, lighter elements (daughter products). Numbers along lines shows half-life for each element.

In figure 1, it may be seen that uranium-238 decays to thorium-234, thorium-234 decays to protactinium-238 and protactinium decays to uranium-234, etc. The amount of time needed for one-half of each element (like uranium-238) to change into its daughter product (like thorium-234) is known as its half-life and is indicated by the number written along each connecting line between elements. Thus it takes 4.5 billion years (by) for one-half of the uranium-238 to change (decay) to thorium-234.

The change from one element to another element in figure 1 involves the process of radiation. Alpha particles, beta particles and gamma particles are all part of this radiation process. Alpha particles are of special interest because they are large, energetic particles that cause molecular damage to human tissue. Alpha particles are involved in the decay of uranium-238, uranium-234, lead-214, thorium-230, radium-226, radon-222, and polonium-218, 214 and 210.

SIGNIFICANCE OF RADON GASES

Radon-222 is an inert gas and moves easily in soil, water and air without chemically combining with other elements. However, the distance it may travel is limited by its short half-life of 3.8 days. Radon from soil tends to be concentrated in the basement or lower levels of a building thus demonstrating the short distance of transportation. Water may transport radon greater distances because of the great mobility of water. For example, water from a radon enriched well may travel through pipes to an upstairs shower, where the radon will be released into the air.

Radon gas is easily inhaled into the lungs. However, being inert, it does not chemically combine with other substances in the lungs, and is exhaled very quickly. Lung tissue damage occurs mainly through alpha radiation. Repeated unhalalation and exhalation of radon-enriched air is thought to produce lung cancer in some individuals. Additional lung damage occurs through decay of radon-222 into lighter daughter products like polonium-218, polonium-214, and polonium-210. Polonium is chemically active and can be retained for a considerable amount of time in lung tissue, and much tissue damage may occur.

ARKANSAS GEOLOGY AND RADON GAS

The natural decay of uranium-238 ultimately produces radon gas. Although uranium-rich minerals are present in many rock types, certain rock types are noted for unusual concentrations or uranium. Igneous rocks like granite and rhyolite, sedimentary rocks like sandstone, shale and phosphorate, and metamorphic rocks, like slate, phyllite, schist, and granite gneiss all may contain locally abnormal amounts or uranium.

Arkansas contains some of the above mentioned rock types, some of which are locally enriched by uranium. A good example is the Devonian age Chattanooga Shale Formation in Benton County, Northwest Arkansas. The shale of this formation is dark gray to black and locally enriched in organic material. Uranium tends to be concentrated in shale in which there is an unusual amount of organic material present. Thus the vertical and horizontal extent of radioactive shale is restricted by the limited extent of the contained organic material. In the 1960's, the author discovered the restricted extent of radioactive rock in the Chattanooga Formation when evaluating the uranium potential of northwestern Arkansas.

A phosphate-rich rock present at the very top of the Chattanooga Formation is also unusually radioactive. This rock, probably a phosphorate, is very thin to absent in many places, but increases in thickness in some localities. Thus radioactive material in the Chattanooga Formation is associated with both black shale and phosphorate. The Geologic Map of Arkansas, published in 1976, shows the general distribution of the Chattanooga Formation in Northwestern Arkansas.

TESTING FOR RADON GAS

The Environmental Protection Agency (EPA) suggests testing for radon gas be conducted in two stages. The first test is of short duration, several minutes to several weeks, and is used to determine whether a building has a high radon gas content. If the air in the building tests less than 4 piococurries of radon per liter of air then no further testing is recommended. Short duration tests are best performed in the late Fall or early Spring when the heating-cooling system is not operating. The test should be conducted in the lowest level of the building in an area where the test monitor will not be disturbed. Consistent test results exceeding 4 piococurries per liter require further long duration testing. Any test result exceeding 200 piococurries per liter requires occupants to take immediate action (abandon building). This secondary test normally lasts from several months to one year. Long duration testing gives a better idea of variations in radon content over the test period. Radon gas content of the air in a building may be 60% higher during winter months than during summer months.

TEST KITS

Many brands of test kits are now available for home use. These kits fall into two general categories:

A. Activated Charcoal Monitors These kits are for short duration testing (less than one week). The kit should be returned to the manufacturer for analysis.

B. Alpha-Track Detectors These kits are also called track-etch detectors, and are used for long duration testing (one to twelve months). Upon completion of the test the detector should be returned to the manufacturer for analysis.

SUMMARY

Radon gas in the air you breath may be detrimental to your health in that a link between radon and lung cancer seems to exist. This gas and some of its harmful daughter products is always being liberated from crustal rock through radioactive decay. Buildings constructed on rock or soil containing radon should be tested for the presence of more than 4 picocurries per liter of radon in the air within the structure. A positive test requires corrective measures be taken as soon as possible by a competent building contractor. The cost of such action is usually not great.

APPENDIX B

MINERAL RESOURCES OF ARKANSAS
by
RONALD H. KONIG

INTRODUCTION

Mineral resources are widespread in Arkansas, and include metallic ores, non-metallic or industrial minerals, and fuels (combustible materials). Metallic ore deposits are confined mostly to north-central Arkansas in the Ozark Mountains and to west-central Arkansas in the Ouachita Mountains. A much wider distribution is associated with industrial minerals, in that nearly all 76 counties show some annual production. Counties with little or no annual production are under populated and have a marginal transportation system. Fuel production is confined mostly to northwestern and southern Arkansas.

An excellent source of detailed information concerning the distribution and production of natural resources in Arkansas up to 1966 was published as Bulletin 645 by the U.S. Bureau of Mines in 1969, under the title "Natural Resources and Industries of Arkansas." Although this bulletin is out of print, it is available in many libraries throughout Arkansas and continues to be the best source for information of this kind.

METAL ORE RESOURCES

A surprisingly great number of metal ores have been produced in Arkansas over the years. Included in the list are aluminum, manganese, vanadium, lead, zinc, and mercury, along with minor quantities of antimony, titanium, copper, and iron. Figure 1 shows the general distribution of some of these resources.

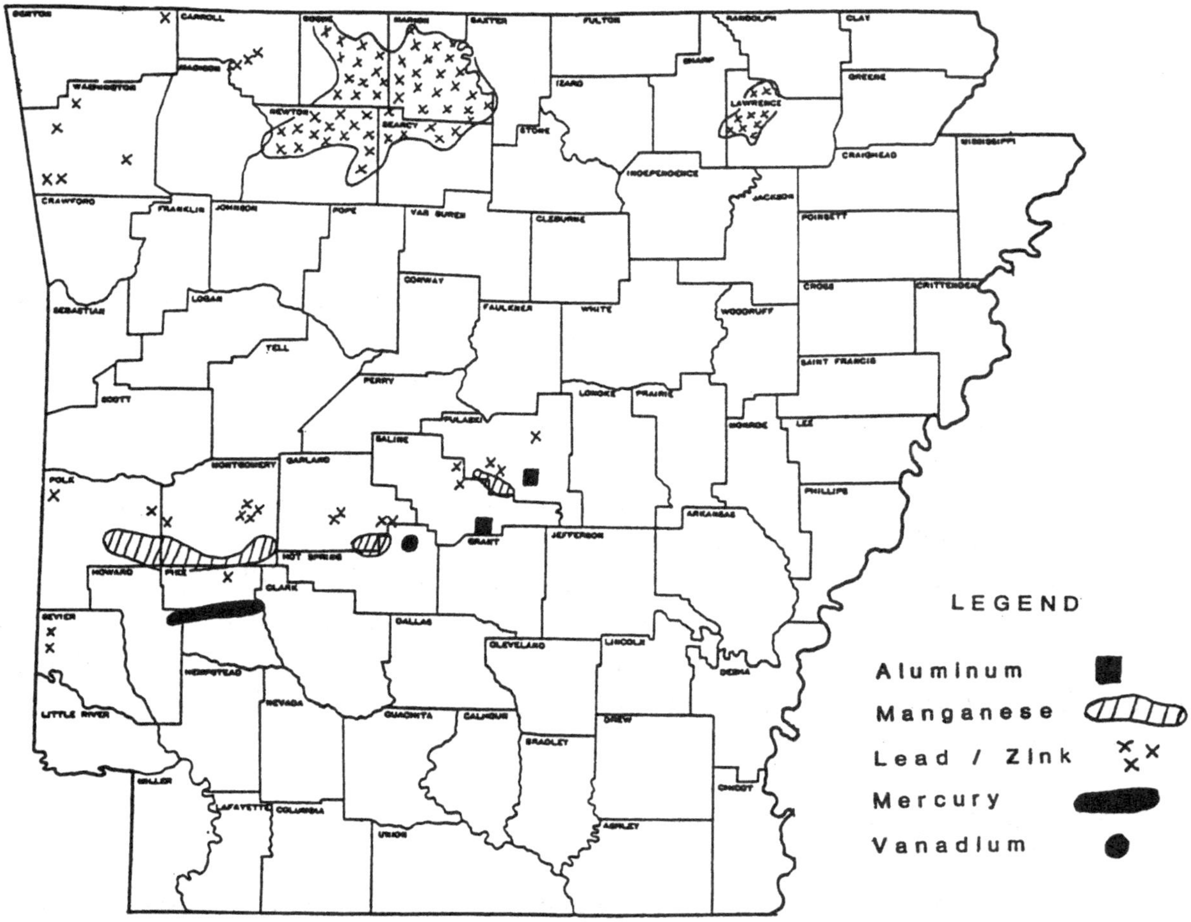

Figure 1. Outline map showing the general distribution of metal ore resources in Arkansas (after U.S.B.M. Bulletin 645, 1969).

Geological investigations have shown that nearly all metal ore resources in the State have resulted from the circulation of heated fluids through ancient rock. Lead and zinc resources in north-central Arkansas were deposited from warm water moving through fractured limestone and dolostone of Ordovician to Mississippian age. Galena (lead sulphide) and sphalerite (zinc, iron sulphide) are the major minerals in these deposits. These same ore minerals, accompanied by small amounts of chalcopyrite (copper, iron sulphide), formed in Paleozoic age rock of the Ouachita Mountains from similar solutions. The ore minerals occur as fracture fillings and as disseminated grains in both areas.

Manganese ores are present in the Batesville area (northeastern Arkansas) and in the west-central part of the Ouachita Mountains. In the Batesville area the manganese ore minerals include cryptomelane, psilomelane, pyrolusite, hausmannite, lithiophorite, and several others, all of which are basically manganese oxides. These ore minerals are confined to Ordovician age rocks assigned to the Fernvale Limestone and Cason Shale formations. Similar manganese minerals are present in the west-central part of the Ouachita Mountains. Here manganese is confined mostly to Devonian-Mississippian age Arkansas Novaculite Formation and to a lesser extent the Mississippian age Ten Mile Creek Formation. In both cases the minerals fill irregular fractures and parallel bedding planes.

Aluminum is obtained from the sedimentary rock called bauxite. The bauxite resulted from the weathering of Cretaceous age syenite exposed in the area between Hot Springs and Little Rock. This weathering converted aluminum-rich feldspar in the syenite into an aluminum-rich soil known as laterite. Tropical weathering started about 50 million years ago and continued for 15 million years to about 35 million years ago (Figure 2). The lateritic soil was slowly buried by younger sediment and changed into a sedimentary rock called bauxite by the process of lithification. Bauxite was discovered in Arkansas by John B. Branner, former State Geologist, in 1887. Bauxite production began in 1896, with the product used only as a road surfacing material. By 1899, however, bauxite was being converted into the metal aluminum, and Arkansas production began to increase from about 5,000 long tons of ore in 1899 to over 2,000,000 long tons of ore by the time production was gradually phased out in the 1970's. Although considerable bauxite remains in the area, present economic conditions do not justify additional production.

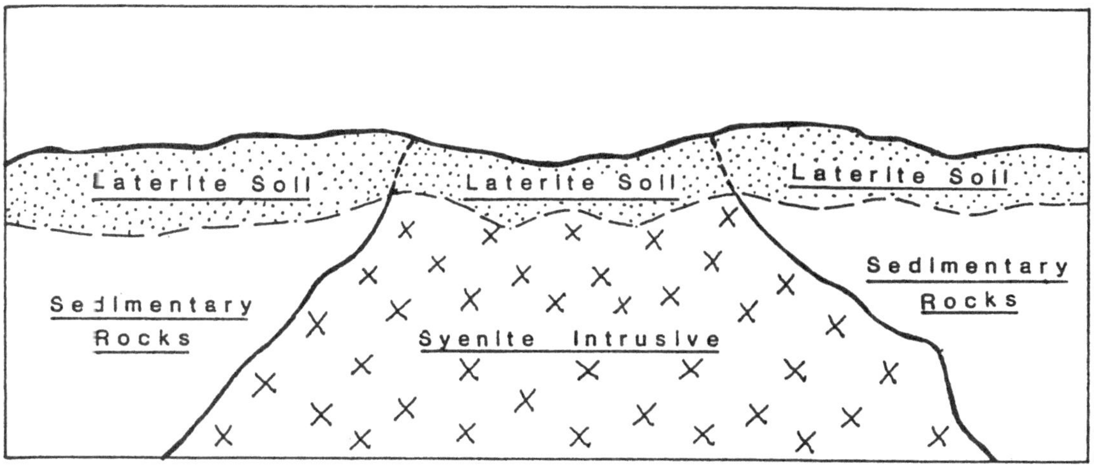

Figure 2. Cross sectional diagram showing tropical weathering conditions producing a lateritic soil from both syenite and adjacent sedimentary rock.

Vanadium ore has been produced since 1965 from soil and very decomposed rock within and adjacent to syenite intrusions known as the Potash Sulphur Springs complex (figure 1). This Cretaceous age igneous complex, which is located in Garland County about seven miles southeast of Hot Springs, was intruded into the Arkansas Novaculite Formation. Mineralizing solutions invaded along the contact between the syenite and the Arkansas Novaculite and ultimately formed vanadium-rich minerals like montrosite (vanadium oxide) and secondary vanadium minerals like fervanite and hewettite. These deposits are the only ones in the United States ever mined exclusively for vanadium. Most vanadium is obtained as a byproduct of uranium production.

FUEL RESOURCES

Major fuel resources or minerals in Arkansas include petroleum (oil), natural gas, bituminous coal and lignite. These resources are confined mostly to the Arkansas River valley and Southern Arkansas. Collectively these resources have contributed more to the State's mineral wealth than all other mineral resources combined. Figure 3 shows the general distribution of these oil and gas resources.

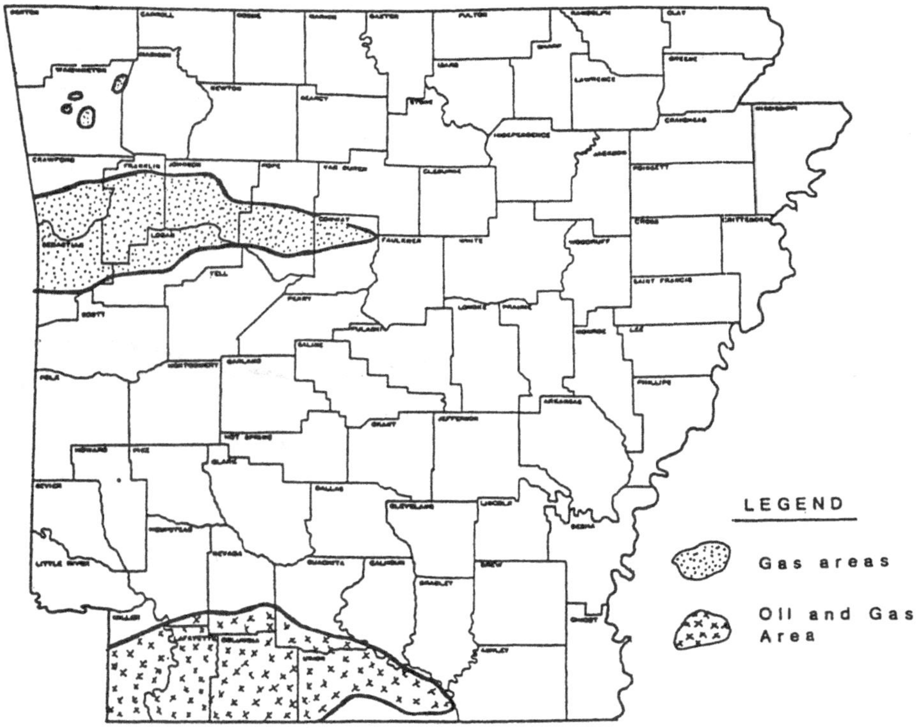

Figure 3. Outline map showing the general distribution of major oil and gas resources in Arkansas (after U.S.B.M. Bulletin 645, 1969).

Petroleum (oil) was first produced at a profit in Union County in southern Arkansas in January 1921. This oil well led to the development of the well known El Dorado field that has produced close to 100 million barrels of oil to date. Since the discovery of the El Dorado field, over 150 additional oil fields have been established in southern Arkansas. Although some of these fields have been abandoned over the years, many others remain active. Most active wells are located within an area about 110 miles from west to east and 40 miles from north to south. Hydrocarbon production in this area is from rocks that are either Jurassic or Cretaceous in age. The Cretaceous age Ozan Formation is the largest producer and the Smackover Formation, of Jurassic age, is the second largest producer. Structural highs known as anticlines serve a traps for much of the oil, but local sand lenses and bars also contain considerable oil.

Natural Gas is produced in two separate areas in the State (figure 3). Wet gas, always associated with crude oil, was discovered near El Dorado in southern Arkansas in 1921 upon the drilling of the first commercial oil well in the State. It is interesting to note however, that dry natural gas (not associated with crude oil) was first encountered in shallow wells drilled near Fort Smith back in 1887. Commercial production of this dry gas did not begin until 1902. Exploration for both types of natural gas has increased over the years due to the population growth and development of both the Arkansas River Valley and farther north in the Arkansas plateau region. There are over 70 known fields in this area today. Each year gas production declines in some fields (some may even be abandoned) and new ones are discovered. Production from these fields comes from sandstone that has been folded and faulted. Anticlinal traps have provided most of the dry gas produced to date.

Wet gas (gas with liquid hydrocarbons) is confined to an area south of the Ouachita Mountains in southern Arkansas. This gas is always associated with crude oil and therefore, everything that was discussed previously under the heading of petroleum is directly related to wet gas distribution and production.

Bituminous coal was first extracted in significant amounts about 1870 and has exceeded 2,000,000 short tons per year on numerous occasions. All Arkansas coal started out as organic accumulation known as peat about 300 million years ago. The peat was gradually buried by younger sand, silt, and clay deposits and first was transformed into lignite then into bituminous coal by the increased pressures and temperatures associated with the formation of the Ouachita Mountains located to the south. There are at least 19 coalbeds in the Arkansas coalfield which covers an area about 33 miles wide and 60 miles long in Crawford, Franklin, Johnson, Logan, Pope, Sebastian, and Scott counties (figure 4). Only the Lower and Upper Hartshorne coalbed of the McCalister Formation and the Charleston and Paris coalbeds of the Savanna Formation have produced commercial amounts of coal. All of these coalbeds are in the lower part of the Pennsylvanian Period. The Lower Hartshorne coal is more than 14 inches thick over an area of approximately 740 square miles and reaches 8 feet in thickness west of Huntington in Sebastian

County. Upper Hartshorne coal is 14 inches thick over an area of only 28 square miles. The Charleston coal reaches a maximum thickness of 23 inches in several places and is over 14 inches thick in an area over 52 square miles. Paris coal ranges from 14 to 32 inches in thickness, but covers an area of only 18 square miles. These coalbeds have been mined by both underground methods and surface strip mining. Abandoned strip mines remain today as testimony to past mining operations. Today strip mining is closely regulated to prevent abandonment of unsightly operations.

Lignite is widespread in both southern and northeastern Arkansas. Tertiary age sedimentary rocks assigned to the Wilcox and Claiborne groups host all commercial lignite deposits in these areas. The lignite began as accumulations of peat in numerous freshwater swamps, in adjacent brackish swamps and in lagoonal areas, all of which developed on a broad coastal plain. Burial of the peat eventually led to the formation of lignite, which is the lowest rank of coal. Arkansas lignite has a moisture content of about 36% which keeps its heating value below 8300 Btu/lb. It also contains 18% incombustible material (ash) and 0.6% sulphur. Figure 4 shows the general distribution of all lignite that is less than 150 feet below the surface in Arkansas. Recent drilling has demonstrated proven reserves of at least 2 billion short tons of lignite. An additional 9 billion tons of lignite is assumed to be present based upon reasonable geological deductions. Since 1973 several major companies have leased about 200,000 acres of land containing commercial quantities of lignite. This lignite will probably be utilized as a fuel for nearby electrical generating plants.

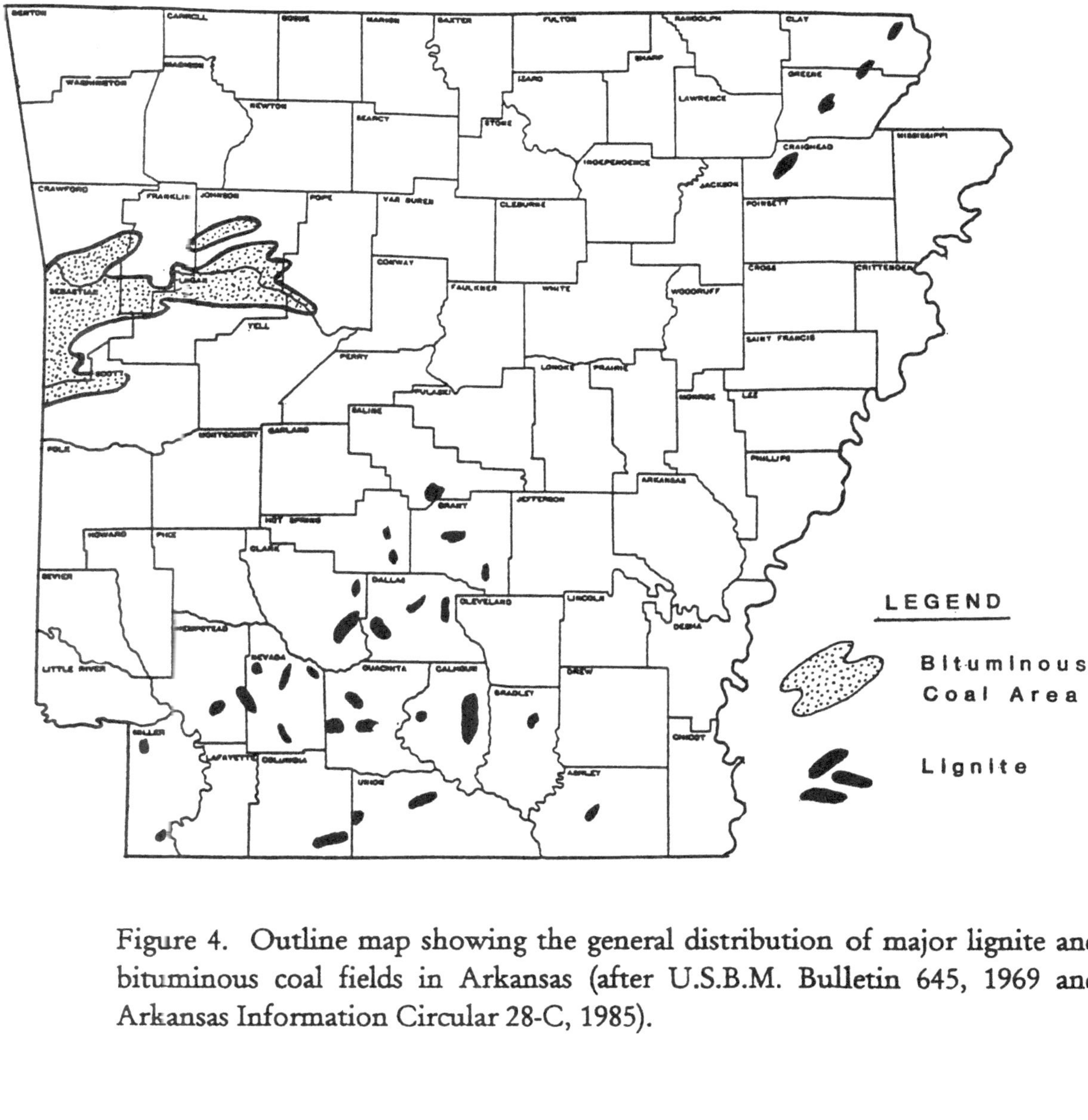

Figure 4. Outline map showing the general distribution of major lignite and bituminous coal fields in Arkansas (after U.S.B.M. Bulletin 645, 1969 and Arkansas Information Circular 28-C, 1985).

NON METALLIC OR INDUSTRIAL MINERALS

Non metallic or industrial minerals, as they are commonly called, include such things as barite, cement, clay, crushed stone, lime, sand, gravel, and glass sand. In Arkansas these minerals are of considerable economic value. Other materials of lesser importance include diamonds, quartz crystals, novaculite, soapstone, phosphate rock, Fuller's earth and tripoli. Figure 5 shows the general distribution of these except for clay, crushed stone, sand, gravel, lime, novaculite, and Fuller's earth. Materials not shown on figure 5 are widely distributed in the State.

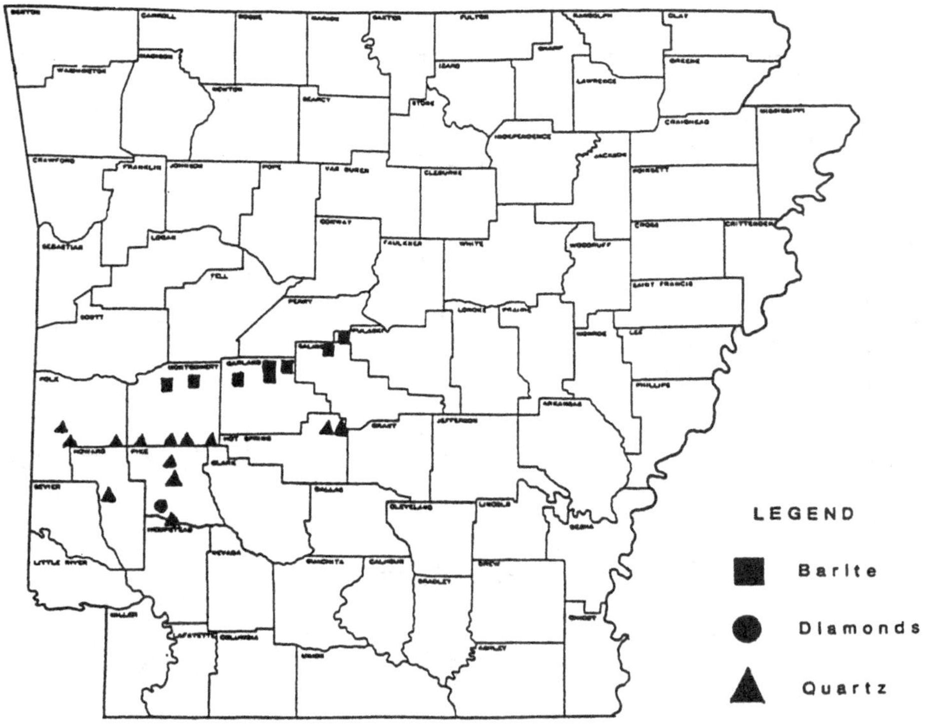

Figure 5. Outline map showing the general distribution of selected industrial minerals in Arkansas (after U.S.B.M., Bulletin 645, 1969).

Barite deposits are primarily confined to black shales in the lower part of the Mississippian age Stanley Formation. Although this formation is exposed throughout the Ouachita Mountains, most deposits occur in the abandoned Chamberlain Creek area in Hot Springs County , and at Fancy Hill and Pigeon Roost Mountain in Montgomery county. Other deposits occur near the town of Hartfield (Polk County) in the middle part of the Arkansas Novaculite Formation, and near the Town of Dierks (Howard County) in Cretaceous sands and gravels (figure 5). The Dierks deposits seem to be of limited importance.

The mineral barite (barium sulphate) is disseminated throughout in the rock in amounts approaching 85 weight percent. Although barite was discovered in Arkansas in 1888, it was not produced in significant quantity until 1940. Arkansas barite production is linked directly to activity in the petroleum industry because all Arkansas barite has been used in the manufacturing of drilling muds.

Cement is a manufactured product consisting mostly of limestone (calcium carbonate) and lesser amounts of quartz (silica), clay (aluminum oxide), and iron. This mixture is heated to 1400 to 2700 degrees F (calcined) to drive off moisture, etc., and then ground to a fine powder. The product is then sold as either natural cement or Portland cement, depending upon its composition and degree of calcining. Most cement production in the State is confined to Howard and Little Rock counties in southwestern Arkansas. Chalk from the Cretaceous age Annona Formation, and shale, quartz sand and iron-rich sand from the overlying Nacatoch Formation, provide most of the natural ingredients for the cement. Many limestones and associated sandstones and shales of northern Arkansas could support a large cement industry under proper economic conditions.

Crushed Stone in Arkansas comes from the grinding of chalk (a type of limestone), dolostone, limestone, novaculite, sandstone, slate, and syenite (a type of igneous rock). Grinding reduces the rock to smaller pieces that are sized by a screening process. The product is then used as roadstone, concrete aggregate, riprap for stabilization projects, roofing granules, and railroad ballast. Crushed stone is one of the most important industrial mineral products of the State.

Sand and gravel occurs in abundance in the majority of Arkansas counties. Although many of these deposits supply material for local markets, the increasing population and improving transportation system have resulted in a much wider distribution for Arkansas sand and gravel. In industry, the term sand refers to rock and mineral fragments with a diameter of 0.74 millimeters to 4.76 millimeters and gravel refers to particles with diameters between 4.76 millimeters and 85.0 millimeters. Neither of these grain size ranges agree with grain size ranges for sand and gravel as used in geology. Nearly all sand in Arkansas is composed of quartz grains (silica), whereas gravel may be composed of many kinds of resistant rock. Resistant rocks found in gravel deposits include limestone, dolostone, sandstone, siltstone, chert, novaculite, and slate. Most sand and gravel deposits occur as stream deposits in and adjacent to stream beds. Others are the result of landslide action (talus deposits) and weathering resulting in residual deposits.

Diamonds were discovered in 1906 in an area two to three miles south and southeast of Murfreesboro in Pike County (figure 5). Most diamonds occur in Cretaceous age volcanic pipes (the volcanic cones have been removed by erosion) that are filled with either massive kimberlite (lamproite) or fractured Kimberlite known as volcanic breccia. Four volcanic pipe areas are known and include the Prairie Creek area, the Kimberlite area, the American area, and the Black Lick area. The majority of the diamonds discovered to date have come from weathered volcanic breccia in the Prairie Creek area. Weathering has produced a 2 to 3 foot thick surface material called "black ground". This altered material gives way to unaltered volcanic breccia or kimberlite below. Diamonds are recovered from this weathered or altered igneous rock. Diamonds discovered to date are white, yellow, and brown and range up to 40.23 carats in weight. The Uncle Sam diamond (40.23 carats) was discovered in 1924. Other large stones include the Star of Murfreesboro (34 carats) discovered in 1964, and the amarillo Star (16 carats) discovered more recently by a visitor to the area. The Arkansas Department of Parks and Tourism purchased the Prairie Creek area in 1972 and allows the public to search for diamonds for a minimal entrance fee. Over 50,000 diamonds have been discovered to date.

Quartz crystals occur in many areas of the Ouachita Mountains, but are particularly abundant in Montgomery and Garland Counties (figure 5). Quartz veins, which usually trend parallel to the layering in the enclosing sandstones and shales, may range in thickness from a few inches to as much as 25 feet. Quality quartz crystals which are essentially pure silica, come from residual clay deposits in openings in the Blakley and Crystal Mountain formations. Individual crystals are removed by hand and range in size from a fraction of an inch to as much as 10 feet in length. Most of the crystals are milky colored, with only a very small proportion being clear. Early settlers in Arkansas used well formed crystals for both jewelry and as ornamental mineral specimens. During World War II an increased demand for quartz crystals used for frequency control in radios and radar devices, resulted in increased quartz mining in Arkansas. Arkansas quartz is also used in optical equipment, fiber optics, as seed crystals (lasca) for growing synthetic quartz crystals under laboratory conditions and in the manufacturing of fusing quartz which has many chemical, thermal and electrical uses. Arkansas is known worldwide for its quality museum quartz specimens.

APPENDIX C

EARTHQUAKES IN ARKANSAS
by
RONALD H. KONIG

INTRODUCTION

An earthquake is the trembling or shaking of the ground caused by the release of energy stored in rock within the earth. This stored energy or stress is released when the rock breaks or faults (undergoes strain), much in the same way as energy is released when a stick breaks after being bent too far. When the rock breaks seismic waves radiate away from the place of faulting and cause the ground to shake. This phenomenon occurs over one million times each year in the earth. Fortunately the vast majority of earthquakes are confined to zones where the lithosphere is being broken in response to the slow movement of rock in convection cells operating below the lithosphere. These zones of breakage define the edges of lithospheric plates. Some plates move apart, some move together and others slide past each other. Thus the distribution of earthquakes, like volcanos, is explained to a great extent by the Theory of Plate Tectonics.

EARTHQUAKES IN ARKANSAS

Earthquakes in Arkansas are generally confined to three distinct areas. One earthquake belt coincides with the northern edge of the Ouachita Mountains. A second belt lies along and generally south of the Ouachita Mountains in southern Arkansas. Finally, a third area of pronounced earthquake activity is located in northeastern Arkansas. This belt also involves eastern Tennessee, southern Missouri, southeastern Illinois and western Kentucky (figure 1).

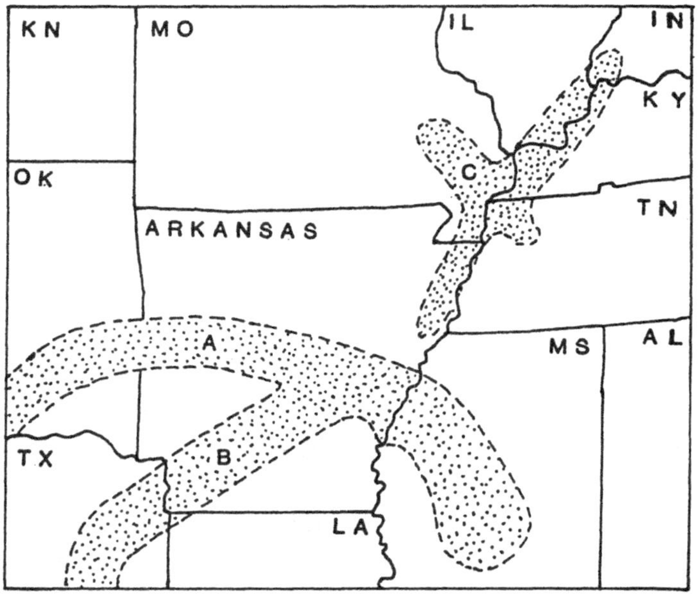

Figure 1. Seismic regions of Arkansas showing Ouachita belt (A), southern Arkansas belt (B) and northeastern belt (C). Modified from K.C. Jackson, Arkansas Geological Commission Information Circular 26, 1979.

The size of earthquakes in Arkansas has been reported by means of two scales. The Modified Mercalli Scale (1931) measures the size of an earthquake on a scale of 1 to 12 with respect to damage reported by individuals who experienced the earthquake. An earthquake assigned a number 1 on this scale is hardly felt by individuals in the area, whereas a number 5 is felt by nearly everyone although there is little or no damage to buildings. A number 7 earthquake is felt by everyone and causes minor damage to buildings of good construction and considerable damage to poorly constructed building. Earthquakes assigned a number 12 cause total destruction. In contrast, the Ritcher Scale (1932) estimates the total amount of energy released by a particular earthquake on a scale of 0 to 9. These magnitude numbers are obtained from actual earthquake records (seismograms) by measuring the height of individual waves on the seismogram. A number 4 earthquake on the Ritcher Scale would appear as a wave on the seismogram with an amplitude of 10 millimeters, provided that the earthquake center was 100 kilometers away from the recording instrument (seismometer).

Seismic waves may travel through both solid rock and unconsolidated rock material (overburden) in Arkansas. Solid rock transmits seismic waves in a very normal manner. Overburden however, tends to amplify the same seismic waves, resulting in a more complex and intense motion of the ground. Structures with foundations resting on solid rock may receive little or no damage, whereas similar structures resting upon unconsolidated rock material may be heavily damaged by an earthquake of the same strength.

Beneath northeastern Arkansas there is a major fracture zone that trends northeast generally parallel to the Mississippi River valley. This fracture zone which has been active on and off since late Precambrian time, is known as the Realfoot graben or rift. At the end of Precambrian time (545 million years ago) this feature was an elongate valley which eventually filled with sediment. More recent movement (activity) along this fracture zone provided the Mississippi River with a convenient place to flow and deposit about three thousand feet of very young, unconsolidated sediment (figure 2). Many buildings, roads, pipelines, etc. now rest upon or within this unconsolidated material.

The New Madrid Earthquake Zone coincides with the Realfoot graben and extends generally from Marked Tree, Arkansas Northeast through New Madrid, Missouri to Cairo, Illinois, a distance of about 150 miles. This area has been seismically active throughout recorded history. In 1811-1812 there was a series of unusually strong earthquakes reported from this area. Over 1800 earthquakes were reported by individuals living in this area during a six month interval, three of which were rather noteworthy. The first of these three earthquakes occurred in Arkansas on December 16, 1811 and may have had a surface magnitude of about 8.6 on the Ritcher Scale. This estimate of magnitude is based upon reported damage and not upon observation of actual seismic records, as the seismograph was not invented until around 1897. A second earthquake (8.4 magnitude) occurred on January 23, 1812 in southeastern Missouri. It was followed by 15 major aftershock earthquakes. Finally, the third major quake occurred a little farther north in southeastern Missouri with an estimated surface magnitude of 8.7.

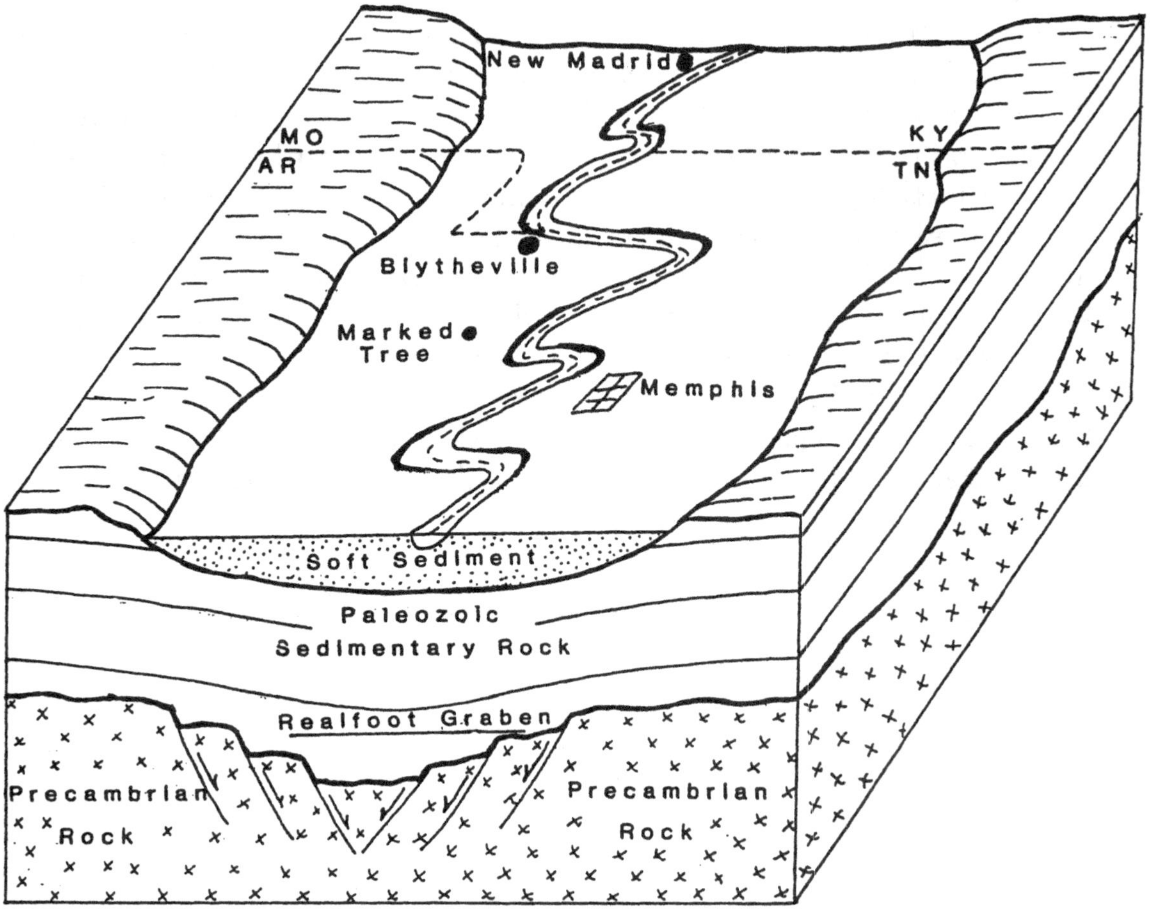

Figure 2. Block diagrams showing the relationship of the Realfoot graben to younger sediment and the Mississippi River valley.

The three earthquakes described above caused much surface damage, but very little loss of life due to a low population density in the area. Only one person is known to have died during these events, and reportedly she ran until exhausted and died of fright. Physical damage consisted of landslides, uplift and subsidence, fissuring, and sand and water extrusion. So strong were these earthquakes that church bells rang as far away as Boston. Buildings in the area were not extensively damaged because they were mostly single story wood frame homes.

Recent geological studies suggest that we should expect other large earthquakes in the future along the New Madrid zone. It is impossible to predict when the next "Big One" will occur, but it is virtually certain that at least a moderate-sized earthquake will affect the area within the next 50 or so years. Even a moderate-sized earthquake is expected to cause extensive physical damage to populated areas. Failed buildings will surely result in many deaths. Greatly affected areas will lie within the Mississippi River valley where the surface material consists of unconsolidated sand, silt and clay.

An earthquake study was conducted by Geologists with the U.S. Geological Survey in the early 1980s at the request of the Federal Emergency Management Agency (FEMA). Results of this study were published in 1983 as CUSEPP Reports, 81-1 and 82-3. These reports concluded that many areas in and immediately adjacent to the New Madrid fault zone would suffer extensive damage as a result of an earthquake with similar magnitude to those associated with the 1811-1812 events. Figure 3 shows anticipated damage levels, as related to the Modified Mercalli Scale, for the New Madrid fault zone and adjacent areas. Fayetteville lies approximately on the number 7 line. However Fayetteville rests upon solid rock (bedrock) and would therefore not experience as much damage as indicated by the number 7 rating. Instead, all residents would feel the earthquake and there would be minor damage to poorly constructed buildings. Well constructed buildings would not be damaged.

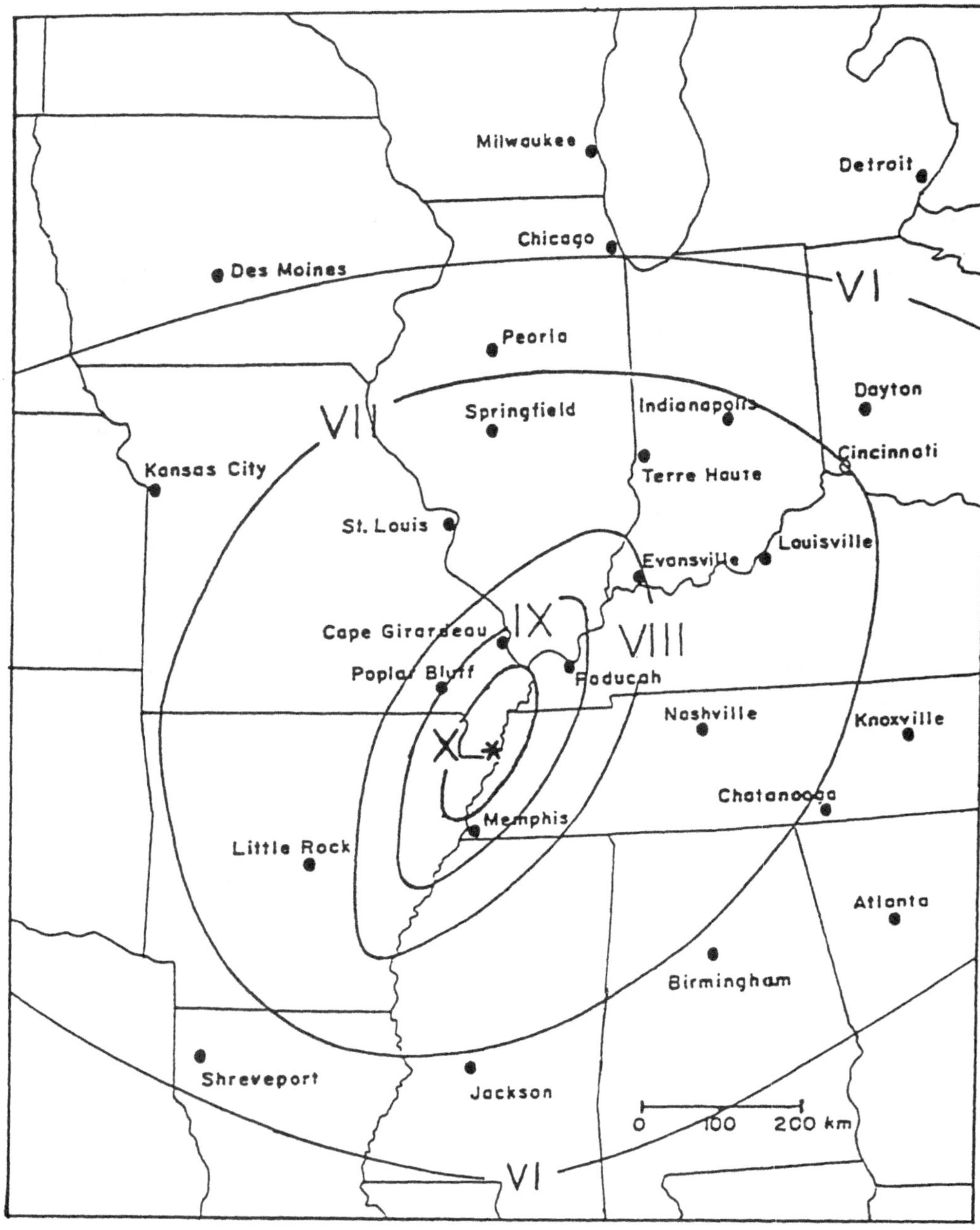

Figure 3. Map showing lines of equal seismic intensity (Mercalli Scale) for an 1811-1812 sized earthquake assumed to occur toward the northeastern end of the New Madrid fault zone. For areas underlain by unconsolidated material, add 1 or 2 units to the intensity value shown. Subtract 1 or 2 units from intensity values shown for areas where bedrock is exposed to the surface. Modified from CUSEPP Report number 81-1, 1981.